FUNDAMENTALS OF
LASER DIODE AMPLIFIERS

FUNDAMENTALS OF LASER DIODE AMPLIFIERS

H. Ghafouri-Shiraz
University of Birmingham, UK

JOHN WILEY & SONS
Chichester · New York · Brisbane · Toronto · Singapore

Other Wiley Editorial Offices

John Wiley & Sons, Inc., 605 Third Avenue,
New York, NY 10158-0012, USA

Jacaranda Wiley Ltd, 33 Park Road, Milton,
Queensland 4064, Australia

John Wiley & Sons (Canada) Ltd, 22 Worcester Road,
Rexdale, Ontario M9W 1L1, Canada

John Wiley & Sons (SEA) Pte Ltd, 37 Jalan Pemimpin #05-04,
Block B, Union Industrial Building, Singapore 2057

Library of Congress Cataloging-in-Publication Data

Ghafouri-Shiraz, H.
     Fundamentals of laser diode amplifiers / H. Ghafouri-Shiraz.
        p.   cm.
     Includes bibliographical references and index.
     ISBN 0 471 95872 7
     1. Semiconductor lasers.   2. Optical amplifiers.   I. Title.
TA1700.G53   1995
621.36′6 — dc20   95-32238
                                                       CIP

British Library Cataloguing in Publication Data

A catalogue record for this book is available from the British Library

ISBN 0 471 95872 7

Typeset in 10/12pt Times by Dobbie Typesetting Ltd
Printed and bound in Great Britain by Bookcraft (Bath) Ltd
This book is printed on acid-free paper responsibly manufactured from sustainable forestation,
for which at least two trees are planted for each one used for paper production.

This book is dedicated to

My Father, Haji Mansour, for the uncompromising principles that guided his life.

My Mother, Rahmat, for leading her children into intellectual pursuits.

My Supervisor: The Late Professor Takanori Okoshi, for his continuous guidance, encouragement, inspiring discussion and moral support. A distinguished scientist and a great teacher who made me aware of the immense potential of optical fibre communications.

My Wife, Maryam, for her magnificent devotion to her family. My constant companion and best friend, she has demonstrated incredible patience and understanding during the rather painful process of writing this book while maintaining a most pleasant, cheerful and comforting home.

My Children, Elham, Ahmad-Reza and Iman, for making everything worthwhile.

To all of my Research and Undergraduate Students, for their excellent and fruitful research work, and for many stimulating discussions, which encouraged and motivated me to write this book.

CONTENTS

Preface   xi

Acknowledgements   xv

1 Introduction   **1**
   References   7

2 Basic Principles of Optical Amplifiers   **9**
   2.1 Introduction   9
   2.2 Interaction of Radiation with a Two-Level System   10
        2.2.1 Radiative processes   10
        2.2.2 Spontaneous emission   11
        2.2.3 Stimulated emission   12
        2.2.4 Absorption   12
        2.2.5 Optical gain   14
   2.3 Characterisation of Optical Amplifiers   16
        2.3.1 Signal gain   16
        2.3.2 Frequency bandwidth   18
        2.3.3 Saturation output power   19
        2.3.4 Noise figure   20
   2.4 Ideal Optical Amplifiers   22
   2.5 Practical Optical Amplifiers   23
        2.5.1 Performance limits of the amplifier signal gain   24
        2.5.2 Performance limits of the amplifier bandwidth   24
        2.5.3 Performance limits of the saturation output power   24
        2.5.4 Performance limits of the noise figure   25
   2.6 Summary   29
   2.7 References   29

3 Optical Amplification in Semiconductor Laser Diodes   **33**
   3.1 Introduction   33
   3.2 Principles of Optical Amplification in Semiconductor Lasers   33

      3.2.1   Optical processes in semiconductors   34
      3.2.2   Analysis of the optical gain in semiconductors   37
  3.3   Semiconductor Laser Diodes as Optical Amplifiers   44
      3.3.1   Optical amplification using homojunctions   44
      3.3.2   Optical amplification using heterostructures   46
  3.4   Types of Semiconductor Laser Amplifiers   48
      3.4.1   Operational classification   48
      3.4.2   Structural classification   51
  3.5   Radiative Transition in Semiconductors   53
      3.5.1   Stimulated emissions   53
      3.5.2   Spontaneous emissions   55
  3.6   Applications of Semiconductor Laser Amplifiers   56
      3.6.1   Non-regenerative repeaters   56
      3.6.2   Pre-amplifiers to optical receivers   58
      3.6.3   Bistable and switching applications   59
      3.6.4   Other applications   61
  3.7   References   62

4 Analysis of Transverse Modal Fields in Semiconductor Laser Amplifiers (SLAs)   **67**
  4.1   Introduction   67
  4.2   Solution of Transverse Modal Fields in Rectangular Optical Waveguides   68
      4.2.1   Solution for a three-layer slab (planar optical waveguide)   68
      4.2.2   Solution for a rectangular dielectric waveguide using modal field approximations   75
      4.2.3   Application of Effective Index Method (EIM) for calculating propagation constants for transverse modal fields in rectangular dielectric waveguides   78
      4.2.4   Other methods to solve for transverse modal fields and the dispersion characteristics of rectangular dielectric waveguides   82
  4.3   Applications of Solutions of Transverse Modal Fields in SLAs   83
      4.3.1   Analysis of the modal gain coefficients   84
      4.3.2   Design of a polarisation insensitive Travelling Wave Amplifier (TWA)   87
  4.4   Importance of Transverse Modal Field Properties in SLAs   98
  4.5   References   99

5 Analysis and Modelling of Semiconductor Laser Amplifiers: Gain and Saturation Characteristics   **103**
  5.1   Introduction   103
  5.2   Analysis of Semiconductor Laser Amplifiers with a Uniform Gain Profile   104
      5.2.1   Amplifier gain formulation in semiconductor laser amplifiers   105
      5.2.2   Gain saturation formulation in semiconductor laser amplifiers   110
      5.2.3   Appraisal on using a uniform gain profile in analysing SLAs   112
  5.3   General Analysis of Semiconductor Laser Amplifiers (A Brief Review)   113
      5.3.1   Analysis using rate equations   113
      5.3.2   Analysis using travelling-wave equations   114
  5.4   Analysis of Semiconductor Laser Amplifiers Using Transfer Matrices   117
      5.4.1   A brief review of matrix methods   117

	5.4.2	Analysis of longitudinal travelling fields in SLAs using the Transfer Matrix Method (TMM)	120
	5.4.3	Analysis of SLAs with a non-uniform gain profile using the Transfer Matrix Method (TMM)	125
	5.4.4	Computational considerations	127
5.5	An Equivalent Circuit Model for SLAs		130
5.6	Applications		135
	5.6.1	Structural effects on amplifier gain	135
	5.6.2	System considerations	139
5.7	Analysis of Gain Saturation in a SLA With a Uniform Material Gain Profile		141
5.8	Summary		143
5.9	References		144

6 Analysis and Modelling of Semiconductor Laser Amplifiers: Noise Characteristics   **149**

6.1	Introduction		149
6.2	Formulation of Noise in Semiconductor Laser Amplifiers		150
	6.2.1	Photon statistics formulation	150
	6.2.2	Rate equation approach	157
	6.2.3	Travelling-wave equations formulation	159
6.3	Analysis of Noise in SLAs Using the Equivalent Circuit Model		159
	6.3.1	Representation of spontaneous emissions in a SLA by an equivalent circuit	160
	6.3.2	Validity of modelling spontaneous emissions by an equivalent circuit	162
	6.3.3	Effects of stray reflections on the spontaneous emission power from a SLA	170
6.4	Applications		175
	6.4.1	Device design criteria	176
	6.4.2	System considerations	182
6.5	Analysis of SLA Spontaneous Emission Power Using the Green Function Approach		183
	6.5.1	Travelling-wave amplifier (TWA)	184
	6.5.2	Fabry–Perot amplifiers	185
6.6	Summary		186
6.7	References		187

7 Experimental Studies on Semiconductor Laser Amplifiers   **191**

7.1	Introduction		191
7.2	Basic Set-up for Measurements		191
	7.2.1	The laser source	192
	7.2.2	Semiconductor laser amplifier	199
	7.2.3	Detection circuit	199
7.3	Experimental Studies on Recombination Mechanisms		200
	7.3.1	Principles of the experimental measurement	201
	7.3.2	Experimental procedures	202
	7.3.3	Results and discussions	202
7.4	Measurement of Gain Characteristics		204
	7.4.1	Experimental set-up	204
	7.4.2	Experimental procedures	209
	7.4.3	Results and discussions	211

7.5   Measurement of Noise Characteristics   214
    7.5.1   Experimental set-up   214
    7.5.2   Experimental procedures   215
    7.5.3   Results and discussions   219
7.6   Summary   221
7.7   References   221

8   Conclusions   **223**
8.1   Summary of the Book   223
8.2   Limitations of the Research Study   226
    8.2.1   Limitations on theoretical studies   226
    8.2.2   Limitations on experimental studies   227
    8.2.3   Limitation on the scope of this book   227
8.3   Future Work   227
8.4   References   228

PREFACE

In the latest decade we have witnessed a tremendous advance in telecommunications technology. With the rapid growth and sophistication of digital technology and computers, communication systems have become more versatile and powerful. This has given a modern communication engineer two key problems to solve: (i) how to handle the ever-increasing demand for capacity and speed in communication systems, and (ii) how to tackle the need to integrate a wide range of computers and data sources so as to form a highly integrated communication network with a global coverage.

The foundations of communication theory show that by increasing the frequency of the carrier used in the system, both the speed and the capacity of the system can be enhanced. This is especially true for modern digital communication systems. As the speeds of computers have increased dramatically over recent years, digital communication systems operating at a speed which can match these computers have become increasingly important. Rather than the electronic circuitry, it is now apparent that the upper bound on the speed of a communication system is limited by the transmission medium. An example that illustrates this is the 80486 PC, which can perform computations with a clock speed of 66 MHz or higher. However, the speed of modems connecting these PCs have just recently reached 1.44 MHz. This is 45 times slower than the electronic clock in the PC. One of the reasons for such a mismatch is that modems use telephone lines (which are typically twisted-pair transmission lines) and these cannot operate at very high frequencies. To improve the speed and hence capacity of the system, we not only need to switch to a carrier with a higher frequency, but to switch to an alternative transmission medium.

Given the preceding argument, the reader will not be surprised by the rapid development of optical communications during the past 20 years. Ever since Kao and his co-workers discovered the possibility of transmitting signals using light in circular dielectric waveguides, research in optical communication systems has developed at an unprecedented pace and scale. Optical communications offer two distinct advantages over conventional cable or wireless systems. First, because the carrier frequency of light is in the region of THz (i.e. 10^{14} Hz), it is possible to carry many more channels than radio waves or even microwave systems. Secondly, the former advantage can be realised because of the development of a matching transmission medium, namely optical waveguides (including fibres and planar structures). Optical waveguides not only provide the necessary frequency bandwidth to accommodate a potentially large number of channels (and hence a huge capacity), but also offer an immunity from electromagnetic interference from which the traditional transmission medium often suffers.

In addition to optical waveguides, another key area of technological development, which plays a crucial role in the success of optical communication systems, is optical devices. The rapid growth of semiconductor laser diodes has allowed optical transmitters to be miniaturised and become more powerful and efficient. Both the fabrication and theoretical research in semiconductor lasers have given rise to a wide range of components for optical communication systems. For example, from conventional buried heterostructure laser diodes to the recent development of multiple quantum-well lasers and from simple Fabry–Perot structures to (i) distributed feedback (DFB) structures, (ii) single cavity laser diodes and (iii) multiple cavity laser diodes. Laser diodes are not only important in compact disc players, but they also provide coherent light sources which are crucial in enhancing the speed and range of transmission of optical communication systems.

The technological forces that gave us optical waveguides and semiconductor laser diodes, have recently explored theoretical research and manufacturing technology to develop other innovative devices that are crucial in optical communications, for example optical amplifiers, optical switches and optical modulators. Previously *optical/electronic* conversion devices had to be used to perform these functions, but the bandwidth of these was limited. The integration of semiconductor laser diodes with optical waveguide technology allows such components to be developed specifically for optical communications. This force of integration does not stop here. The advent of photonic integrated circuits (PICs), which are ICs built entirely with optical components, such as laser diodes, waveguides and modulators will further enhance the power and future prospects of optical communication networks.

In view of the increasing pace of development and growing importance of optical communication technology, I believe that students, researchers and practicing engineers should be well equipped with the necessary theoretical foundations for this technology, as well as acquiring the necessary skills in applying this basic theory to a wide range of applications in optical communications. There are of course many good books about optical communication systems, but they seldom direct their readers to concentrate on the two key aspects behind the success in optical communications which we have discussed above. I am attempting to fill this gap with this book. I will be concentrating on the basic theory of optical waveguides and semiconductor laser technology, and I will illustrate how these two aspects are closely related to each other. In particular, I will examine how semiconductor laser amplifiers have been developed based on applications of the basic theory of these two areas.

Throughout this book it is intended that the reader gains both a basic understanding of optical amplification and a factual knowledge of the subject based on device analysis and application examples. I hope that this book will be beneficial to students aiming to study optical amplification, and to the active researchers at the cutting edge of this technology.

This book is organised as follows: Chapter 1 explores the state of the art of optical fibre communication systems in this rapidly evolving field. A short introduction includes the historical development, the principles and applications of semiconductor laser amplifiers in optical fibre communications, the general optical system and the major advantages provided by this technology. In Chapter 2 the basic principles and important performance characteristics of optical amplifiers will be outlined. Chapter 3 gives an introduction to optical amplification in semiconductor laser diodes. Chapters 4–6 deal with the analysis of semiconductor laser amplifiers (SLAs). In these chapters the waveguiding properties and the basic performance characteristics of SLAs (i.e. amplifier gain, gain saturation and noise) will be studied. Also a new technique, which is based on an equivalent circuit model, will be introduced for the analysis of SLAs. The implications of SLAs for optical fibre communication system performance will also be discussed. In Chapter 7 the accuracy and limitations of the equivalent circuit model will be investigated by comparing both theoretical and experimental results for actual devices. Finally, Chapter 8 is devoted to some concluding remarks and comments.

The book is referenced throughout by extensive end-of-chapter references which provide a guide to further reading and indicate a source for those equations and/or expressions that have been quoted without derivation.

The principal readers of this book are expected to be undergraduate and postgraduate students who wish to consolidate their knowledge in lightwave technology, and also researchers and practicing engineers who need to equip themselves with the foundations to understand and use the continuing innovations in optical communication technologies. The reader is expected to be equipped with a basic knowledge of communication theory, electromagnetism and semiconductor physics.

Finally, I must emphasise that optical communications is a rapidly growing technology with very active research. After reading the book I hope that the reader will be equipped with the necessary skills to apply the most up-to-date technology in optical communications.

Dr H. Ghafouri-Shiraz
December 1995, Birmingham, UK

ACKNOWLEDGEMENTS

I owe particular debts of gratitude to my former research student, Dr C. Y. J. Chu, for his excellent research work on semiconductor laser diode amplifiers. I am also very grateful indeed for the many useful comments and suggestions provided by reviewers which have resulted in significant improvements to this book. Thanks also must be given to the authors of numerous papers, articles and books which I have referenced while preparing this book, and especially to those authors and publishers who have kindly granted permission for the reproduction of some diagrams. I am also grateful to both my undergraduate and research students who have helped me in my investigations. Finally, my special thanks go to Jack Henderson, a former senior lecturer at the University of Birmingham, who kindly reviewed the manuscript. Thanks also to Mr M. S. Rodger for his excellent drawings.

Dr H. Ghafouri-Shiraz
December 1995, Birmingham, UK

INTRODUCTION

The demand for high-capacity, long-haul telecommunication systems is increasing at a steady rate, and is expected to accelerate in the next decade [1]. At the same time, communication networks that cover long distances and serve large areas with a large information capacity are also in increasing demand [2]. To satisfy the requirements on long distances, the communication channel must have a very low loss. On the other hand, a large information capacity can only be achieved with a wide system bandwidth which can support a high data bit rate (> Gbit/s) [3]. Reducing the loss whilst increasing the bandwidth of the communication channels is therefore essential for future telecommunications systems.

Of the many different types of communication channels available, optical fibres have proved to be the most promising [4, 5]. The first advantage of an optical fibre is its low attenuation. Typical values of attenuation factor in modified chemical vapour deposition (MCVD) optical fibres are plotted against wavelength of the electromagnetic carrier in Figure 1.1 [6]. At present, optical fibres with loss coefficients of less than 0.25 dB/km around emission wavelengths of 1.55μm are available [7]. This remarkable progress in fibre manufacturing technology has led to wide applications of long-distance optical fibre communications in recent years. Furthermore, optical fibres can also transmit signals over a wide bandwidth because the electromagnetic carrier in optical fibres has a frequency in the optical frequency region ($\approx 10^{14}$ Hz). Hence, optical fibres can also carry many baseband channels each with a bandwidth of the order of GHz using wavelength division multiplexing (WDM) [8, 9]. For these reasons, optical fibre communication systems have attracted a lot of attention in recent years, and much research has been carried out to optimise their performance.

Figures 1.2(a)–(d), respectively, show the properties of various elements used in optical fibre communication systems, namely the main materials and wavelengths used for different light sources, optical detectors and optical amplifiers where there have been rapid advances recently. With semiconductor optical amplifiers, by changing the crystal composition the wavelength band (i.e. amplifiable waveband) can be selected as required from short to long

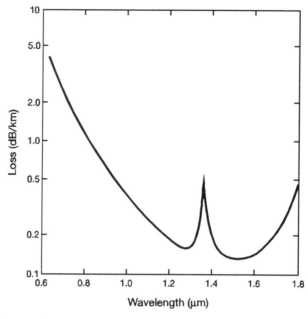

Figure 1.1 Attenuation coefficient of MCVD optical fibres as a function of emission wavelength

wavelengths (see Figure 1.2(c)). Furthermore if a travelling wave device is used, broad band operation over 10 THz or so is possible. Rare-earth-doped optical fibre amplifiers, on the other hand, have an amplifiable waveband that is essentially determined by the dopant material, and in the 1.55 μm band this is limited to erbium. Erbium doping is therefore of great practical value since it allows fabrication of a fibre amplifier suitable for operation at 1.55 μm, which is the waveband of lowest loss in silica optical fibres.

A typical configuration for an optical fibre communication system is shown in Figure 1.3. The optical fibre acts as a low loss, wide bandwidth transmission channel. A light source is required to emit light signals which are modulated by the signal data. To enhance the performance of the system, a spectrally pure light source is required. Advances in semiconductor laser technology, especially after the invention of double heterostructures (DHs), resulted in stable, efficient, small-sized and compact semiconductor laser diodes (SLDs) [10–12]. Using such coherent light sources increases the bandwidth of the signal which can be transmitted in a simple intensity modulated (IM) system [13]. Other modulation methods, such as phase-shift keying (PSK) and frequency-shift keying (FSK), can also be used [4, 14]. These can be achieved either by directly modulating the injection current to the SLD or by using an external electro- or acousto-optic modulator [11, 15].

The modulated light signals can be detected in two ways. A direct detection system as shown in Figure 1.3 employs a single photo-detector [13, 16] which acts as a square law detector as in envelope detection in conventional communication systems [3]. Although such detection schemes have the inherent

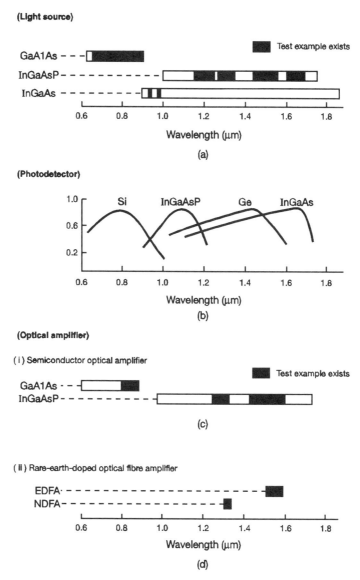

Figure 1.2 Wavebands of components used in optical fibre communication systems (after [28])

advantage of simplicity, the sensitivity of the receiver is limited [17]. In order to detect data transmitted across the optical fibre with a higher bit rate, the signal-to-noise ratio at the input to the receiver must be made as high as possible. In a system without repeaters, this will limit the maximum transmission span of the system [1]. An alternative detection method is to use coherent detection [4, 18], as shown in Figure 1.4. By mixing the signal with a local oscillator at the input to the detector, it can be shown that a higher sensitivity can be achieved if the receiver is designed properly [5]. The principle is similar to that in a heterodyne

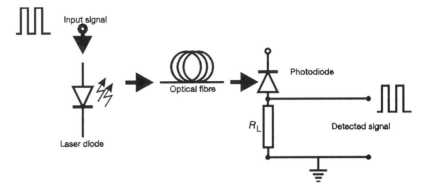

Figure 1.3 Configuration for an optical fibre communication system employing direct detection

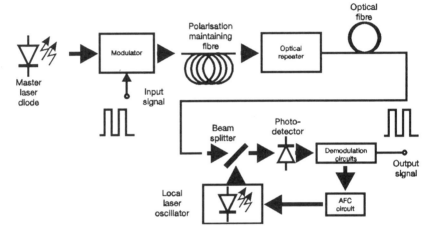

Figure 1.4 Configuration for a coherent heterodyne optical fibre communication system

radio [3]. In this system, one can detect WDM transmission easily by tuning the local oscillator wavelength as in a heterodyne radio system. In practice, however, because of the finite spectral width of the master and/or local oscillators, which are usually SLDs, the limited tunability in SLDs and the extreme sensitivity of the receiver to the states of polarisation of the light signal will severely limit the performance of such complicated receivers [4]. Some of the recent field trials employing coherent detection are shown in Table 1.1.

Although, in theory, coherent detection seems to offer a better performance for optical communications over direct detection, receivers employing this technique are very much at the research stage and their performance has yet to be improved [5]. On the other hand, many existing practical optical communication systems employ direct detection with intensity modulation. In order to use them for the transmission of data with a higher data rate in the future, it is more economical if we can simply improve the input signal-to-noise

Table 1.1 Recent coherent optical heterodyne transmission field experiments

Laboratory	Transmission speed	Modulation scheme	Route	Year
KDD	565 Mb/s	FSK	Submarine cable	1988
BTRL	565 Mb/s	DPSK	Cambridge–Bedford	1988
AT&T	1.7 Gb/s	FSK	Roaring Creek–Sunbury	1989
BTRL	622 Mb/s	FSK, DPSK	Edinburgh–Newcastle	1989
BTRL	620 Mb/s	DPSK	U.K.–Guernsey	1989
NTT	2.5 Gb/s	FSK	Matsuyama–Ohita-Kure	1990

ratio of the optical receiver instead of replacing or upgrading existing components in the systems like using new optical fibres or replacing the entire receiver using coherent detection with a new modulation scheme. In addition, the problem of retrieving WDM signals using direct detection has been overcome by using tunable optical filters, which are cheaper than tunable SLDs, at the input of the receiver [1]. Hence, it appears that, if the input signal-to-noise ratio of the receivers can be improved, existing direct detection systems with intensity modulation can be used for transmissions with an even higher data rate.

The weak signal at the receiver in many optical communication systems arises because of the accumulation of losses along the optical fibres [1]. Although the loss can be as low as $0.2\,dB/km$ for optical fibres operating around $1.55\,\mu m$, for a long transmission span this can build up to a significant loss, which will degrade signal power and hence the overall system performance [19]. Two ways to improve the signal-to-noise ratio of an optical receiver are possible: we can either boost the optical signal power along the transmission path using in-line repeaters [20], or boost the optical signal power at the input of the receiver by a pre-amplifier [21]. For many applications, both methods must be used to improve the system performance. In-line repeaters can be constructed using electronic circuits, which consist of photodetectors, electronic circuits for the demodulation of the signals, amplification circuits for loss compensation and laser diode driving circuits for regeneration. These conventional electronic repeaters are known commonly as regenerative repeaters. With them, the signal-to-noise ratio at the input of the receiver can indeed be improved. However, since the specification and subsequent design and configuration of this type of regenerator depends heavily on the modulation format, data bit rate, multiplexing scheme and, in the case of optical networks, the number of branches emerging from a node, they are uneconomical because of their poor flexibility [22].

To solve the flexibility problem for in-line repeaters and to provide a pre-amplifier for optical receivers, we must be able to amplify light signals directly. For in-line repeaters, direct optical amplification avoids regeneration circuits in the repeaters, can be used for any modulation format of the signal [23] and provides a maximum flexibility for applications in systems [22]. Repeaters employing such techniques are commonly known as non-regenerative

Table 1.2   Recent transmission experiments with (a) fibre amplifiers and (b) semiconductor laser amplifiers

Year	Laboratory	Bit rate (Gbit/s)	Distance (km)	Comments
(a) Fibre amplifiers				
1989	NTT	1.8	212	Booster + pre-amplifier used
1989	BTRL	0.565	—	DPSK system
1989	KDD	1.2	267	Two amplifiers used
1989	NTT	20	—	Soliton transmission
1989	Bell Core	11	260	Two amplifiers used
1989	Fujitsu	12	100	—
1989	KDD	1.2	904	12 amplifiers used
1990	NTT	2.5	2223	25 amplifiers used
1991	BTRL	2.5	10^4	Recirculating loop
1991	NTT	10	10^6	Soliton transmission, 12 amplifiers used
1991	NTT	20	500	Soliton transmission
1992	NTT	2.4	309	4 repeaters + pre-amplifier
1992	NTT	10	309	4 repeaters + pre-amplifier
(b) Semiconductor laser amplifiers				
1986	BTRL	0.14	206	2 amplifiers used
1988	AT & T	1	313	4 amplifiers used
1988	AT & T	0.4	372	4 amplifiers + FSK
1988	Bell Core	—	—	20-channel transmission
1989	BTRL	0.565	400	5 amplifiers + DPSK
1989	KDD	2.4	516	10 amplifiers used
1991	KDD	0.14	546	10 amplifiers used

repeaters, and the devices that perform such tasks are called optical amplifiers, or quantum amplifiers [24]. Because stimulated emissions are involved in the amplification process, which are also responsible for oscillations in lasers, these optical amplifiers are usually called laser amplifiers. These optical amplifiers can also be used as pre-amplifiers to receivers to enhance their sensitivities further [25]. An improvement in system performance by using optical amplifiers as in-line repeaters and/or pre-amplifiers to optical receivers has been reported in numerous experiments, some of which are tabulated in Table 1.2 [26].

The future prospects of long-distance optical communication systems thus depend heavily on the availability of low-cost optical amplifiers which can compensate for the build-up of losses in optical fibre cables over long distances [2, 4]. Two types of optical amplifier exist: semiconductor laser amplifiers (SLAs) and fibre amplifiers (FAs). SLAs are essentially laser diodes operating in the linear amplification region below the oscillation threshold [23, 27], whereas FAs are optical fibres doped with erbium ions (Er^{+3}) to provide optical gain [19]. SLAs have the inherent advantage of compactness and the possibility of integration with other opto-electronic components, whereas FAs have the advantages of easy and efficient coupling with optical fibres. The design and analysis of both these types of optical amplifiers are therefore crucial for future development in optical fibre communication systems.

In this book the principles of semiconductor laser diode amplifiers will be explored. In Chapter 2 the fundamentals and important performance characteristics of optical amplifiers will be outlined. An introduction to optical amplification in semiconductor lasers will be described in Chapter 3. A formal treatment of the analysis of semiconductor laser amplifiers will be given in Chapters 4–6, where the waveguiding properties, and the basic performance characteristics such as gain, gain saturation and noise will be studied. A new technique for analysing SLAs using an equivalent circuit model will also be introduced. Implications for system performance will also be discussed. In Chapter 7 the accuracy and limitations of this model will be investigated by comparing theoretical predictions with the results of experimental measurements on actual devices. The final chapter of this book covers the concluding remarks and comments.

REFERENCES

[1] C. Rolland, L. E. Tarof and A. Somani, "Multigigabit networks: the challenge", *IEEE LTS Mag.*, **3**, pp. 16–26, 1992.

[2] M. G. Davis and R. F. O'Dowd, "A new large-signal dynamic model for multielectrode DFB lasers based on the transfer matrix method", *IEEE Photon. Technol. Lett.*, **4**, No. 8, pp. 838–840, 1992.

[3] H. Taub and D. L. Schilling, *Principles of Communication Systems*, 2nd edition, McGraw-Hill, New York, 1986.

[4] T. Kimura, "Coherent optical fibre transmission", *IEEE J. Lightwave Technol.*, **LT-5**, No. 4, pp. 414–428, 1987.

[5] R. E. Wagner and R. A. Linke, "Heterodyne lightwave systems: moving towards commercial use", *IEEE LCS Mag.*, **1**, No. 4, pp. 28–35, 1990.

[6] M. J. Adams, *An Introduction to Optical Waveguides*, John Wiley and Sons, New York, 1981.

[7] T. Miya, Y. Teramuna, Y. Hosuka and T. Miyashita, "Ultimate low-loss single mode fibre at 1.55 mm", *Electron. Lett.*, **15**, No. 4, pp. 106–108, 1979.

[8] E. Dietrich, B. Enning, G. Grosskopf, L. Kuller, R. Ludwig, R. Molt, E. Patzak and H. G. Weber, "Semiconductor laser optical amplifiers for multichannel coherent optical transmission", *J. Lightwave Technol.*, **7**, No. 12, pp. 1941–1955, 1989.

[9] W. I. Way, C. Zah and T. P. Lee, "Application of travelling-wave laser amplifiers in subcarrier multiplexed lightwave systems", *IEEE Trans. Microwave Theory Technol.*, **38**, No. 5, pp. 534–545, 1989.

[10] G. H. B. Thompson, *Physics of Semiconductor Laser Devices*, John Wiley and Sons, New York, 1980.

[11] A. Yariv, *Optical Electronics*, 3rd edition, Holt-Saunders, 1985.

[12] G. P. Agrawal and N. K. Dutta, *Long-wavelength Semiconductor Lasers*, Van-Nostrand Reinhold, New York, 1986.

[13] J. M. Senior, *Optical Fibre Communications: Principle and Practice*, 2nd edition, Prentice-Hall, New York, 1992.

[14] D. J. Maylon and W. A. Stallard, "565 Mbit/s FSK direct detection system operating with four cascaded photonic amplifiers", *Electron. Lett.*, **25**, No. 8, pp. 495–497, 1989.

[15] A. Yariv and P. Yeh, *Optical Waves in Crystals*, John Wiley and Sons, New York, 1984.

[16] P. E. Green and R. Ramaswami, "Direct detection lightwave systems: why pay more?", *IEEE LCS Mag.*, **1**, No. 4, pp. 36–49, 1990.

[17] H. Kressel, *Semiconductor Devices for Optical Communications*, 2nd edition, Springer-Verlag, Berlin, 1982.

[18] T. Okoshi, "Ultimate performance of heterodyne/coherent optical fibre communication", *IEEE J. Lightwave Technol.*, **LT-4**, No. 10, pp. 1556–1562, 1986.

[19] N. Nakagawa and S. Shimada, "Optical amplifiers in future optical communication systems", *IEEE LCS Mag.*, **1**, No. 4, pp. 57–62, 1990.

[20] S. D. Personick, "Applications for quantum amplifiers in simple digital optical communication systems", *Bell Syst. Technol. J.*, **52**, No. 1, pp. 117–133, 1973.

[21] A. J. Arnaud, "Enhancement of optical receiver sensitivity by amplification of the carrier", *IEEE J. Quantum Electron.*, **QE-4**, No. 11, pp. 893–899, 1968.

[22] H. Nakagawa, K. Aida, K. Aoyama and K. Hohkawa, "Optical amplification in trunk transmission networks", *IEEE LTS Mag.*, **3**, pp. 19–26, 1992.

[23] G. Eisenstein, "Semiconductor optical amplifiers", *IEEE Circuits and Devices Mag.*, pp. 25–30, 1989.

[24] Y. Yamamoto and T. Mukai, "Fundamental of optical amplifiers", *Opt. Quantum Electron.*, **QE-21**, pp. S1–S14, 1989.

[25] Y. Yamamoto and H. Tsuchiya, "Optical receiver sensitivity improvement by a semiconductor laser amplifier", *Electron. Lett.*, **16**, No. 6, pp. 233–235, 1980.

[26] H. Ghafouri-Shiraz, Unpublished work, 1992.

[27] Y. Mukai, Y. Yamamoto and T. Kimura, "Optical amplification by semiconductor lasers", *Semiconductor and Semimetals*, **22**, Part E, pp. 265–319, Academic Press, London, 1985.

[28] S. Shimada and H. Ishio, *Optical Amplifiers and their Applications*, John Wiley and Sons, New York, p. 3, 1994.

2

BASIC PRINCIPLES OF
OPTICAL AMPLIFIERS

2.1 INTRODUCTION

The future prospects of high-speed, long-distance optical fibre communication systems depend heavily on the availability of low-cost optical amplifiers which can compensate for the build-up of losses in optical fibre cables over long distances. Two types of optical amplifiers exist: (i) semiconductor laser diode amplifiers and (ii) fibre amplifiers. Semiconductor laser diode amplifiers are essentially laser diodes operating in the linear amplification region below the oscillation threshold, whereas fibre amplifiers are optical fibres doped with erbium ions (Er^{3+}) to provide optical gain. Semiconductor laser diode amplifiers have the inherent advantages of compactness and the possibility of integration with other opto-electronic components, whereas fibre amplifiers have the advantages of easy and efficient coupling to optical fibres. The design and analysis of both types of optical amplifier are therefore crucial for the future development of optical fibre communications.

To understand fully how optical amplification can be achieved, the interaction between electromagnetic radiation and matter must first be understood. Therefore, in this chapter we will first explore the interaction of radiation with a simple two-level atomic system. This simple model provides the basis for studies of more complex quantum mechanical systems, including those of semiconductors [1].

An understanding of the interaction of radiation with a two-level system enables us to understand the operation of optical amplifiers and from this their fundamental performance characteristics can be derived. In general, such characteristics can be used to describe both fibre amplifiers and semiconductor laser amplifiers [2], so that any optical communication system incorporating either type of optical amplifier can be analysed in a formal and consistent way [3–5]. The performance characteristics of an ideal optical amplifier will be derived after the above discussion. The ideal optical amplifier can be used as a reference against which to assess the ultimate performance of real semiconductor laser amplifiers [6]. Finally, the performance limitations of optical amplifiers, which will determine the ultimate performance of an optical system, will be analysed.

2.2 INTERACTION OF RADIATION WITH A TWO-LEVEL SYSTEM

One way to understand the physics behind optical amplification processes in any optical amplifier is by considering a simple two-level system as shown in Figures 2.1(a)–(c). This description is sufficient to give a fairly accurate qualitative picture of the physical processes that take place inside gas, or solid state, semiconductor lasers or optical amplifiers [1, 7]. There are three fundamental radiative processes that may take place when an electromagnetic wave interacts with a lasing material. These are spontaneous emission, stimulated emission and absorption. Spontaneous emission, because of its very nature, is distributed over a wide range of frequencies. The dynamic behaviour of a laser or an optical amplifier is often described with reasonable precision by a set of coupled rate equations involving the three radiative processes. In their simplest form, these are a pair of simultaneous differential equations describing the population inversion and the laser radiation field as functions of time.

A more accurate picture for semiconductor laser amplifiers will be treated in Chapter 3. Meanwhile, we will first examine how electromagnetic radiation interacts with a two-level system, and then we will see how these interactions give rise to optical gain in optical amplifiers.

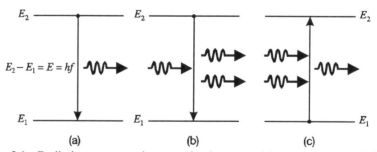

(a)   (b)   (c)

Figure 2.1 Radiative processes in a two-level system: (a) spontaneous emission; (b) stimulated emission; and (c) absorption

2.2.1 Radiative processes

In a two-level optical amplification system the atoms can occupy one of two discrete energy levels, which are separated by a finite energy E. On excitation, the atoms gain energy and transit from the lower energy level 1 to the higher energy level 2. Conversely, the atoms may lose energy and transit from the higher energy level 2 to the lower energy level 1. The transitions between the energy levels of the system can be described accurately by quantum mechanics [1, 7, 8], although a semi-classical treatment can also be used [9, 10]. In the following discussion the latter approach will be used to clarify the analysis. This problem was first analysed by Einstein based on some physically reasonable postulates concerning radiative transitions [9].

Consider an atomic system that has two nondegenerate energy states E_1 and E_2 such that $E = E_2 - E_1$. When an atom undergoes a transition between these two states, it either emits or absorbs a photon of frequency $f = E/h$, where h is Planck's constant [1]. Under normal circumstances the atom will be in the lower level because physical systems tend to be more stable at the lowest energy state. Given the presence of an incoming photon, there is a high probability that an atom will configure itself at the excited energy state, but it returns again to the ground state by giving up a photon of energy E. In what follows we examine the details of the three radiative processes for a proper understanding of the process of achieving population inversion.

2.2.2 *Spontaneous emission*

Spontaneous emission is a random radiative process whereby an electron in the excited state E_2 decays to state E_1 and, in so doing, gives out energy E in the form of a photon, as shown in Figure 2.1(a). If the population density of the excited energy state is N_2 and that of the lower energy state is N_1, then the rate of spontaneous decay of these atoms is found to be proportional to N_2. Hence, we have

$$\left[\frac{dN_2}{dt}\right]_{sp} = -\left[\frac{dN_1}{dt}\right]_{sp} = -A_{21}N_2 \tag{2.1}$$

where A_{21} is called the spontaneous emission probability. The spontaneous emission lifetime $\tau_{21,sp}$, defined as the average time during which the electron survives in the excited state before contributing to a spontaneous emission, is equivalent to $1/A_{21}$. The probability that a particular atom will undergo a spontaneous transition in time dt is $A_{21}dt$. The population density of the excited state is given by

$$N_2 = N_2^0 \exp\left(-\frac{t}{\tau_{21,sp}}\right) \tag{2.2}$$

where N_2^0 is the value of N_2 at $t = 0$. Equation (2.2) justifies the fact that if an atomic system is subjected to only spontaneous emission, N_2 will deplete rapidly thus causing a rise in the population density N_1 of the lower energy state.

2.2.3 Stimulated emission

It is possible that an input photon can interact with an excited system and trigger the emission of excess energy. A photon passing sufficiently close to an excited atom may cause the atom to undergo a radiative transfer before it would otherwise do so spontaneously. The excess photon energy output has the same frequency, phase, direction and polarisation as the stimulating photons. Hence, the process involves an amplification of coherent light [8]. Stimulated emission is associated with a specific energy level difference in a material, and the emitted photons show a high degree of spatial and temporal coherence with the incident photon (see Figure 2.1(b)). The fact that the stimulated and stimulating photons are in phase results in a high degree of coherence in lasers. In practice, a less than perfect temporal coherence is achieved because of the randomness present in radiative transfers due, among other things, to thermal vibrations. In comparison, the spontaneous emission has no definite phase and the resulting photon emission can be in any random direction. The stimulated emission rate is dependent on the incoming field as well as on the number of atoms to be stimulated and is given by

$$\left[\frac{dN_2}{dt}\right]_{st} = -\left[\frac{dN_1}{dt}\right]_{st} = -B_{21}N_2\rho(f) \tag{2.3}$$

where $\rho(f)$ is the average spectral energy density per unit frequency for blackbody radiation [9], and B_{21} is the stimulated emission probability per unit time per unit spectral energy density. Note that $\rho(f)df$ provides the total photon density within the frequency intervals f and df. As will be shown in the next subsection, stimulated emission is the inverse of stimulated absorption.

2.2.4 Absorption

An electron in a lower energy state, after having absorbed an energy equivalent to E, is raised to an excited energy state. In the presence of a quasi-monochromatic radiation field of frequency f, the population of the less excited state is depleted at a rate proportional to both the population of that state and the radiation density. This de-amplification process is called absorption and is shown in Figure 2.1(c). The absorption rate is given by

$$\left[\frac{dN_2}{dt}\right]_{abs} = -\left[\frac{dN_1}{dt}\right]_{abs} = B_{12}N_1\rho(f) \tag{2.4}$$

where B_{12} is the absorption probability per unit spectral energy density. The input energy is accordingly decreased by an amount that was absorbed by the atomic system. At thermal equilibrium, the photon absorption rate must equal the sum of the stimulated and spontaneous emission rates. An atomic system that is simultaneously subjected to all three radiative processes has an overall decay rate given by

$$\frac{dN_2}{dt} = -\frac{dN_1}{dt} = -A_{21}N_2 + B_{12}N_1\rho(f) - B_{21}N_2\rho(f) \tag{2.5}$$

Since at equilibrium $dN_2/dt = dN_1/dt = 0$, using equation (2.5) we can calculate the ratio of the population densities as

$$\frac{N_2}{N_1} = \frac{B_{12}\rho(f)}{A_{21} + B_{21}\rho(f)} \tag{2.6}$$

There would be no absorption and only stimulated emission if N_1 were zero, but there would be only absorption if N_2 were zero. The ratio of the population densities at thermal equilibrium is also given by the Boltzmann distribution

$$\frac{N_2}{N_1} = \frac{g_2}{g_1} \exp(-E/kT) \tag{2.7}$$

where g_i ($i = 1, 2$), referred to as the degeneracy, corresponds to the number of independent ways in which the atom can have the same energy E_i . Note that the ratio N_2/N_1 is always less than unity when $E_2 > E_1$ and $T > 0$ K. Usually, the energy level separation is of the order of 10^{-19} joules or more, and thus at thermal equilibrium (at 300 K) N_1 exceeds N_2 by a factor of the order of 10^8. This implies that optical amplification is not realisable at thermal equilibrium. The spectral energy density $\rho(f)$ can be found by equating equations (2.5) and (2.7) as

$$\rho(f) = \frac{A_{21}}{B_{21}} \left[\frac{g_1 B_{12}}{g_2 B_{21}} \exp(E/kT) - 1 \right]^{-1} \tag{2.8}$$

It may be noted that the spectral energy density is also given by Planck's radiation law as

$$\rho(f) = \frac{8\pi h f^3 n^3}{c^3[\exp(E/kT) - 1]} \tag{2.9}$$

where n is the refractive index of the medium. The only way equations (2.8) and (2.9) can then be equal is if and only if [9, 10]

$$\frac{g_1}{g_2} = \frac{B_{21}}{B_{12}} \tag{2.10}$$

and

$$A_{21} = \left[\frac{8\pi h f^3 n^3}{c^3} \right] B_{21} \tag{2.11}$$

Equations (2.10) and (2.11) are referred to as Einstein's relations. These equations interrelate the coefficients A_{21}, B_{12} and B_{21}. The fact that $B_{12} = B_{21}$, when $g_1 = g_2$, for simplicity, implies that for a given radiation density, stimulated emission and absorption are equally probable. In addition, equation (2.11) confirms that for most applicable frequencies, the rate of spontaneous emission is insignificant when compared with that of stimulated emission. Hence, if favourable circumstances permit, the lasing within a material is expected to be dominated primarily by stimulated emission. Under conditions of thermal equilibrium, however, stimulated emission is not likely. The higher the frequency, the less likely the process. We may conclude that all three radiative processes compete against each other according to equations (2.10) and (2.11). In building a laser or amplifier, we expect to find stimulated emission to be the most dominating radiative process. This scenario can be achieved only by increasing the energy density and by making N_2 larger than $(g_1/g_2)N_1$. This condition is referred to as population inversion.

2.2.5 Optical gain

The previous analysis neglected the spatial dependence of the radiation field in the system. In reality, light is often confined in an optical cavity and interacts along with the atoms which are distributed in the cavity. Hence the spatial dependence of radiation must be taken into account in analysing practical optical amplifiers. To tackle this problem let N_1 and N_2 be the population density of atoms at point z along the propagation direction of radiation for levels 1 and 2, respectively, $A = A_{21}$ and $B = B_{12} = B_{21}$ (i.e. $g_1 = g_2$). Then equation (2.5) may be expressed as

$$\frac{dN_1}{dt} = -\frac{dN_2}{dt} = AN_2 + B(N_2 - N_1)\rho(f, z) \qquad (2.12)$$

Notice that the z-dependence of ρ has been introduced. In general, N_1 and N_2 are functions of both position z and time t. Since each transition of an atom between the two energy levels involves a generation (for spontaneous emission and stimulated emission) or loss (for absorption) of the photon, we can derive a rate equation in terms of photon density S_p at position z and time t as

$$\frac{dS_p}{dt} = AN_2 + B(N_2 - N_1)\,\rho(f, z) \qquad (2.13)$$

Because S_p is also a function of both z and t, we can write it as a total derivative [11], that is

$$\frac{dS_p}{dt} = \frac{\partial S_p}{\partial t} + \frac{\partial S_p}{\partial z}\cdot\frac{dz}{dt} = \frac{\partial S_p}{\partial t} + \frac{\partial S_p}{\partial z}\cdot v_g \qquad (2.14)$$

where $dz/dt = v_g$ is the group velocity of the travelling photons. If the radiation has been interacting with the system for a sufficiently long time such that a steady state has been reached, $\partial S_p/\partial t$ becomes zero and equation (2.14) reduces to

$$v_g\frac{\partial S_p}{\partial z} = AN_2 + B(N_2 - N_1)\,\rho(f, z) \qquad (2.15)$$

Equations (2.12) and (2.15) are the fundamental equations for the analysis of optical amplifiers, which we will examine in more detail in later chapters. Meanwhile, we will concentrate on the solution of equation (2.15). To do this analytically, we assume that the spontaneously emitted photons are scattered and lost instead of being guided along with the incident radiation (i.e. $A = 0$). Denoting the photon density of the guided modes as S_g, the following equation can be obtained:

$$\frac{\partial S_g}{\partial z} = \frac{B(N_2 - N_1)\,\rho(f, z)}{v_g} \qquad (2.16)$$

The photon density of the guided mode and the spectral energy density are not unrelated. In general, the transition frequencies of the atoms even in a simple two-level system possess a finite statistical spread due to finite temperature, and the proportion of atoms involved in the transition with frequencies lying in the range df is given by $F(f)df$, where $F(f)$ is the statistical distribution function describing the spread of transition frequencies of the atoms in the cavity [9]. Hence the photons generated due to radiative transitions will have their frequencies spread in the range df as well. Consider a small section of length dz along the cavity with a cross-section area σ. The spectral energy density $\rho(f, z)$ can be related to S_g by the following equation:

$$\rho(f, z)df = S_g \cdot hf \cdot \sigma \cdot v_g \cdot F(f)df \qquad (2.17)$$

Substituting equation (2.17) into equation (2.16) gives

$$\frac{\partial S_g}{\partial z} = B(N_2 - N_1) \cdot hf \cdot \sigma \cdot S_g \cdot F(f)df \qquad (2.18)$$

Because it is optical power rather than photon density that is usually measured, it is more convenient to express equation (2.18) by the power P over the frequency range df of the guided modes, which is related to the photon density by

$$P(z) = S_g \cdot hf \cdot \sigma \cdot v_g \cdot F(f)df \qquad (2.19)$$

Then equation (2.18) can be re-written in the form

$$\frac{dP}{dz} = hf \cdot \sigma \cdot v_g \cdot B(N_2 - N_1) \cdot P \cdot F(f)df \qquad (2.20)$$

For the frequency range df concerned, equation (2.20) can be solved by integration to give P at $z = z_0$ as

$$P(z_0) = P(0)\,\exp(K(z_0)) \qquad (2.21)$$

where $P(0)$ is the optical power at $z = 0$, and K is known as the *attenuation factor*, which is given by

$$K(z_0) = hf \cdot \sigma \cdot v_g \cdot F(f)df \cdot \int_0^{z_0} B(N_2 - N_1)\,dz \qquad (2.22)$$

It can be seen that the value of K is affected by (i) the population distribution of the atoms in the two energy levels, (ii) the statistical distribution of the transition frequencies $F(f)$, (iii) the exact frequency range df of the measured optical power and (iv) the structural parameters of the cavity (i.e. the parameters σ and v_g). These factors are very important and will be encountered again when discussing amplification in semiconductor lasers in later chapters. At the moment, we will examine (i) in more detail.

Under normal conditions with N_1 and N_2 independent of z (uniformly distributed along the cavity), the distribution of the population density between the two energy levels is described by the following relation [10]:

$$\frac{N_2}{N_1} = \exp(-hf/kT) \qquad (2.23)$$

where k is the Boltzmann constant and T is the temperature of the system measured in kelvin. When T is larger than absolute zero, it can be seen from equation (2.23) that $N_2 < N_1$, and hence K will be negative, as depicted by equation (2.22). This implies that the optical power reduces as the photons travel along z and the radiation is absorbed within the cavity. To provide optical gain (i.e. optical power increasing with z instead of decreasing) the value of *K must be made positive*. As seen from equation (2.22), this can only be achieved if $N_2 > N_1$. Physically, this implies that the stimulated emission rate BN_2 is greater than that of induced absorption BN_1 along the cavity, generating sufficient coherent photons such that the numbers of photons are building up as they travel along the cavity. The condition $N_2 > N_1$ is known as *population inversion*, and the resulting system becomes an optical amplifier [9]. From equation (2.23), it can be seen that this is possible only if T is maintained below absolute zero, and hence this condition is also known as *negative temperature*. According to thermodynamics, this means that energy has to be *pumped* continuously into the system to maintain such population inversion. Therefore, if optical gain has to be introduced into the system, an external energy source has to provide pumping to create a population inversion. In semiconductors such an energy source can take the convenient form of an electrical current [12], whereas in gas and solid state material like fibre amplifiers, the external pumping can take the form of intense light pulses [1].

2.3   CHARACTERISATION OF OPTICAL AMPLIFIERS

After identifying the physical origin of the optical gain in optical amplifiers, we need to investigate the essential characteristics of these devices. Mukai *et al.* [13] identified four important parameters which can be used to describe the performance of any optical amplifier, including semiconductor laser amplifiers (SLAs) and fibre amplifiers (FAs). These parameters are important in defining the performance of optical amplifiers in an optical fibre communication system. They include: the *signal gain*, the *frequency bandwidth*, the *saturation output power* and the *noise figure*.

2.3.1   Signal gain

The most important property of an optical amplifier is its ability to amplify the power of light. As shown in subsection 2.2.5, the optical gain in an optical amplifier is created by external pumping. In SLAs it is provided by the injection of carriers by an electrical current, whereas in FAs which are doped by a rare earth such as erbium (Er^{3+}), the external pumping is provided by a powerful light source [2]. In any case, the gain of the optical amplifier is of primary interest [14], as it determines many other essential factors, like the signal-to-noise (S/N) ratio, when they are incorporated into systems [15]. The signal gain G of an optical amplifier is given, in decibels, by

$$G = 10 \, \log[P_{out}/P_{in}] \tag{2.24}$$

where P_{out} is the light power measured at the output of the optical amplifier, and P_{in} is that measured at the input end [16].

We can further refine the definition of signal gain G by considering the light path. If equation (2.24) describes the input and output light power due to a single light path from input to output of the optical amplifier, the resulting gain is known as the *single pass gain* G_s. If positive feedback is provided (i.e. by reflections from end-facets in semiconductor laser diode amplifiers) as shown by the feedback path in Figure 2.2, the signal gain G becomes

$$G = \frac{G_s}{1 + F_B G_s} \tag{2.25}$$

where F_B is the proportion of output signal which is fed back to the input. Equation (2.25) is analogous to that of an electronic amplifier. When $G = G_s$, it corresponds to amplifiers without feedback ($F_B = 0$). This is true in measuring the gain of SLAs with zero or nearly zero facet reflectivities (known as travelling-wave or near-travelling-wave amplifiers, respectively [17]). If the reflectivities are finite so that some sort of optical feedback is provided to the optical amplifier, then F_B is finite and the resulting amplifier structure becomes a Fabry–Perot amplifier. The physical implications of equation (2.24) and F_B will be discussed again in more detail in Chapter 3 when different types of SLAs are explored. At this point it is sufficient to know how to distinguish between the different meanings of G and G_s.

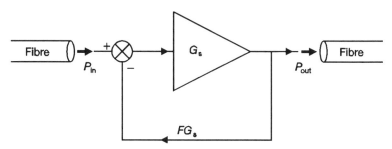

Figure 2.2   An optical amplifier with feedback

Ideally an optical amplifier should have a gain as high as possible. Physically, optical amplifiers with infinite gain are impossible to achieve. An infinite amplifier gain requires an infinite pumping rate to maintain an infinite population inversion, and such an energy source does not exist in nature. Furthermore, according to equation (2.16), an amplifier with infinite gain will actually generate an output without any input. This is equivalent to the condition for oscillation where the amplifier becomes a laser [10]. Therefore the performance limit of the gain in any optical amplifier is that *the amplifier gain is not so large that self-sustained oscillations will be excited*. This is complicated by the fact that stray reflections in the system can provide additional feedback to the amplifier [18, 19], pushing the overall gain of the amplifier towards oscillation for moderate pumping (equation (2.25)). In addition, the maximum signal gain which can be obtained in an optical amplifier is further restricted by gain saturation mechanisms [20]. This subsequently affects the dynamic range of the amplifier, limiting the maximum optical power which can be input to, and output from, the amplifier [21].

2.3.2 Frequency bandwidth

As with any electronic amplifiers, the gain in an optical amplifier is not the same for all frequencies of the input signal. If one measures the gain G for signals with different optical frequencies f, an optical frequency response of the amplifier $G(f)$ can be obtained. This is more commonly known as the optical gain spectrum of the amplifier [22–24]. A typical optical gain spectrum has a finite bandwidth B_0, which is determined by the $-3\,dB$ point in the spectrum as with the conventional definition of the bandwidth in linear systems. The value of B_0 therefore determines the bandwidth of signals which can be transmitted by an optical amplifier, hence affecting the performance of optical communication systems when using them as repeaters or pre-amplifiers [25].

In linear systems a finite bandwidth arises because of the presence of reactive components like capacitors and inductors. Different mechanisms, however, account for the finite bandwidth in optical amplifiers. Again, the exact mechanism varies in different types of optical amplifiers. Nevertheless, two major reasons for finite bandwidth can be identified. The first reason is because of the waveguiding action of the amplifier (or dispersive effects) [26]. In order to amplify light signals they must be confined within the region with optical gain. This can be achieved conveniently in most optical amplifiers by using optical waveguides (e.g. the active region in semiconductor laser amplifiers; see Chapters 3 and 4). Any optical waveguide, whether semiconductor laser diode amplifiers (SLAs) or circular dielectric types (as in fibre amplifiers), possess cut-off frequencies and hence a finite bandwidth [27, 28]. The second reason is that the material gain itself has a finite bandwidth [29–31]. As we have discussed in subsection 2.2.5, in real systems at finite temperature, each energy level of a system which an atom can occupy has a statistical spread according to the principles of thermodynamics and atomic bonding, and hence these two energy levels spread from two distinct levels to form two bands [1, 12]. This allows a range of frequencies to be amplified by the system, resulting in a finite material gain bandwidth. The corresponding mechanisms for the material gain spectra in SLAs will be discussed in more detail in Chapter 3.

The above two reasons can account for the finite bandwidth in optical amplifiers. Usually, both dispersive effects and the finite frequency range of the optical gain will occur in all types of optical amplifiers and the resulting bandwidth of the amplifier is a summation of these two effects. Many proposed models of SLAs have neglected the dispersive effect in deriving the amplifier gain spectra $F(f)$ [11, 32], because experimental work showed that material gain effects dominate in SLAs [33]. However, it can be shown that for many SLA structures the full material gain bandwidth cannot be utilised because of the presence of resonant behaviour in the device, which significantly reduces the actual bandwidth of the amplifier [16]. This will be discussed in further detail in the later chapters.

2.3.3 Saturation output power

The signal gain of an optical amplifier is not only limited to a finite range of optical frequencies, but also by a finite range of input and hence output power, as discussed in subsection 2.3.1. According to equation (2.24), whenever the input power to the optical amplifier is increased, the output power should be increased simultaneously by a scaling factor G. However, in real situations the input power cannot be increased for ever. Experimentally, it is observed that in all optical amplifiers, once the input power is increased to a certain level P_s, the gain G starts to drop [13]. If the measured gain G is plotted against P_{out}, a curve similar to that in Figure 2.3 can be obtained. The output power at the $-3\,dB$ points, as shown in Figure 2.3, is known as the *saturation output power* P_{sat}, and the corresponding *saturation output intensity* I_{sat} can be used to describe the gain saturation effect quantitatively. When the amplifier gain G is measured against the output light intensity I_{out} a similar phenomenon of saturation can

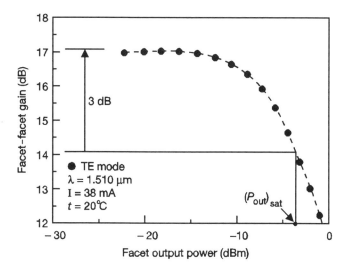

Figure 2.3 A typical experimental result of amplifier gain versus output optical power

be observed, which can be described in an optical amplifier with the gain given by [13].

$$G_s = \exp(g_m L) = \exp[g_0 L/(1 + I_{out}/I_{sat})] \qquad (2.26)$$

where g_m, measured in cm^{-1} is the *material gain coefficient* which is determined by the population inversion level in the material [31], L is the length of the amplifier and g_0 is the unsaturated value of g_m.

This behaviour of optical amplifiers is known as *gain saturation* [34]. A qualitative explanation for this phenomenon can be obtained from the two-level system model. The pumping source creates a fixed amount of population inversion at a particular rate, and on the other hand the amplification process is continuously draining the inverted population by creating stimulated emissions. As we increase the input power, a point arrives where the rate of draining due to amplification is greater than the rate of pumping, such that the population inversion level can no longer be maintained at a constant value and starts to fall. Thus the gain of the system starts to fall, as observed in Figure 2.3. In other words, gain saturation arises simply because of the conservation of energy. Because of this saturation phenomenon, when an optical amplifier is used to amplify several channels of light signals (as in a wavelength division multiplexed system), if the *total* optical power input to the amplifier (*not* the optical power of an individual channel) from all the channels exceeds P_{sat}, the amplifier will be saturated [13]. The saturation behaviour is complicated by the fact that in real optical amplifiers the mechanisms involved are far more complex than in the two-level system described above, and often the non-linear effects of the material gain will also play a role [35].

2.3.4 Noise figure

Optical amplifiers, like other electronic amplifiers, are not free from noise. This aspect of optical amplifiers is different from the previous ones in the sense that the noise process in all optical amplifiers is due to an identical mechanism: spontaneous emission [36, 37]. Because these spontaneous emissions are random events, the phases of the emitted photons are also random. This can be proved using quantum mechanics [8, 9]. If the spontaneous emission photons happen to be emitted close to the direction of travel of the signal photons, they will interact with the signal photons [38], causing both amplitude and phase fluctuations [39, 40]. In addition, these spontaneous emissions will be amplified as they travel across the optical amplifier towards the output [41]. Hence at the output of the amplifier the measured power consists of both the amplified signal power GP_{in} and the amplified spontaneous emissions (ASE) power P_N. This means that

$$P_{out} = GP_{in} + P_N \qquad (2.27)$$

As with electronic amplifiers, a figure of merit can be attributed to an optical amplifier to describe its noise performance. In an electronic amplifier, its noise performance is measured by a *noise figure F*, which describes the degradation of the signal-to-noise (S/N) ratio due to the addition of amplifier noise [42]. Mathematically, it is given as

$$F = \frac{(S/N)_{\text{in}}}{(S/N)_{\text{out}}} \qquad (2.28)$$

where $(S/N)_{\text{in}}$ is the signal-to-noise ratio at the amplifier input, and $(S/N)_{\text{out}}$ is that at the amplifier output. This description can be equally applied to optical amplifiers [6]. It can be shown either by quantum mechanics [6] or semi-classical arguments [43] that, because spontaneous emission is unavoidable in any optical amplifier, the minimum noise figure of an ordinary optical amplifier is 3 dB (unless the optical amplification process is due to parametric amplification [10], for which a noise figure of 0 dB can be achieved [6, 43]). A more detailed analysis can be found in subsection 2.5.4.

Noise in an optical amplifier is the most important parameter. It will not only limit the signal to noise ratio in systems incorporating optical amplifiers [3, 44, 45], but it will also impose other limitations on various applications of optical amplifiers in optical fibre communications. For example, consider several optical amplifiers cascaded in tandem along a transmission span as linear repeaters to compensate fibre losses [15, 46] (Figure 2.4). The *amplified spontaneous emission* (ASE) noise power P_N contributes a part of the output power P_{out} of a particular amplifier in the chain, as described by equation (2.19), and becomes the input to the next amplifier. Therefore P_N can be further amplified by subsequent amplifiers. Because gain saturation depends on

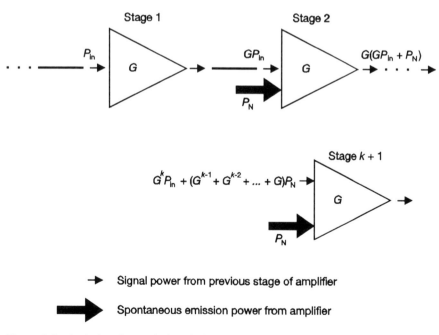

Figure 2.4 A chain of cascaded optical amplifiers and the subsequent propagation of amplified spontaneous emissions

the total amount of power input to the amplifier (subsection 2.3.3), the ASE noise from the output of the earlier stages in the optical amplifier chain can be so large that it will saturate the following ones. If the reflectivities on both input and output ends of the amplifier are low, backwardly emitted ASE from the amplifiers of the later stages can also input to amplifiers of the earlier stages, enhancing saturation due to ASE. With fibre amplifiers, this build-up of ASE noise can actually result in self-sustained oscillations along the transmission span of the fibre if discontinuities, and hence reflections, are present along the path [18, 19]. Even though such back reflections are small, in a long transmission distance involving a relatively large number of optical amplifiers as linear repeaters, sufficient ASE power may build up along the amplifier chain to trigger oscillation. To minimise this effect, optical isolators can be installed along the fibre link to cut off backward-emitted ASE [47], but this will prevent the system being used for bi-directional transmission [48].

In addition to the degradation of performance in terms of power, the phase contamination of the signal due to spontaneous emissions is also manifest as additional amplitude and frequency noise [49], especially due to stray reflections from optical interfaces [18]. As the input signal to optical amplifiers already has a finite amount of phase noise due to the finite spectral spread of the laser source [50, 51], further enhancement of noise from the amplifier is possible. This will further degrade the performance of optical communication systems using phase modulation and coherent detection (e.g. phase-shift keying, PSK).

2.4   IDEAL OPTICAL AMPLIFIERS

The four parameters discussed in the previous sections, namely gain, bandwidth, saturation output power and noise, are used by Mukai *et al.* [13] to characterise an optical amplifier. Other factors such as linearity are important in some specific applications like multi-channel systems [52], but for general discussion these four parameters are sufficient. For instance, when amplifiers are used as linear repeaters, the gain–bandwidth determines how many optical amplifiers are needed along the transmission span of a particular distance [46]. On the other hand, the maximum number of optical amplifiers that can be used as repeaters is limited by the ASE power of each amplifier's output (i.e. their noise figures), as well as their saturation output power [17]. The exact performance requirements of optical amplifiers vary with different system applications. Therefore for a particular system application a specific set of *performance requirements* for the optical amplifiers must be specified [16]. This will be examined in further detail in Chapter 3. It should be noted, however, that all of these four parameters are important in determining the performance in many different applications of optical amplifiers. Therefore it is possible to identify some basic performance requirements that are universal for all major applications. An optical amplifier that satisfies these performance requirements is described as an *ideal* optical amplifier. It is an ideal device in the sense that its performance characteristics are suitable for a very wide range of applications.

As the principal use of optical amplifiers is to amplify light signals, an ideal optical amplifier should have a gain value G as high as possible. A wide optical bandwidth is also desirable, so that the amplifier can amplify a wide range of signal wavelengths. Saturation effects introduce undesired distortion to the output, and hence an ideal optical amplifier should have a very high saturation output power to maximise its dynamic range with minimum distortion. Finally, an ideal amplifier should have a very low noise figure (minimum 3 dB), which will minimise ASE power at the output, thereby maximising the number of optical amplifiers that can be cascaded.

The preceding discussion on the performance requirements for ideal optical amplifiers assumes them to be used as lightwave amplifiers. This type of application includes in-line repeaters and pre-amplifiers to optical receivers [16]. On the other hand, these performance requirements are equally valid for other applications. For instance, since an optical amplifier can be used as a gating optical switch ("on" when pumping is on, "off" when pumping is off), its response time and hence its optical bandwidth is a crucial factor (but note that for switching based on *optical bistability* instead of pumping, this is not true; see Chapter 3 for details), in addition to the requirement of low noise figures for the amplifiers (to avoid spurious switching). Similarly, when the amplifier is used as an optical modulator (by modulating the refractive index of the amplifier which can be achieved very easily in semiconductor laser amplifiers by simply modulating the electrical current [53]) and as a detector [54], its optical bandwidth should be wide, its optical gain should be high and the saturation output power should also be high (to achieve good linearity) with minimum noise. The detailed performance requirements for these various applications will be discussed in Chapter 3, with special reference to semiconductor laser amplifiers.

Of course, real devices fail to meet the above criteria of an ideal optical amplifier. Nevertheless, these ideal characteristics can be seen as targets to which device engineers can aspire. To do so, a more detailed understanding of the operation of optical amplifiers is required.

2.5 PRACTICAL OPTICAL AMPLIFIERS

The preceding discussions on the ideal performance requirements of optical amplifiers were purely qualitative. In practice it is useful to have some quantitative criteria in order to assess how far the actual performance of optical amplifiers falls short of the ideal optical amplifier. Furthermore, in real physical situations, optical amplifiers with infinite gain, bandwidth and saturation output power are impossible to obtain because of the physical limitations of various processes taking place inside the amplifier. Hence, it is more practical to study optical amplifiers that are operating within their performance limits in terms of the following four characteristics [6].

2.5.1 Performance limits of the amplifier signal gain

Physically, optical amplifiers with infinite gain are impossible to achieve as an infinite pumping rate is required to maintain an infinite population inversion [1, 12]. Furthermore, according to equation (2.24), an amplifier with infinite gain will actually generate an output without having any input. This is equivalent to an oscillator and hence the amplifier becomes a laser [10]. This is actually the operational principle of a laser oscillator [1, 12]. Therefore the performance limit of the gain of any optical amplifier is that the amplifier gain is not so large that self-sustained oscillations will be excited. In practice, the maximum signal gain that can be obtained in an optical amplifier is further restricted by gain–saturation mechanisms [55]. This dynamic characteristic of the amplifier can severely reduce the maximum signal gain for high input and output powers [56]. Therefore both gain and saturation output power are limited by gain saturation mechanisms, and are interrelated in this respect [57, 58]. The details of the effects of gain saturation will be discussed again in subsequent chapters with special reference to semiconductor laser diode amplifiers.

2.5.2 Performance limits of the amplifier bandwidth

Concerning the bandwidth, a material that can amplify the whole electro-magnetic spectrum (i.e. has infinite bandwidth) does not exist in nature. In other words, an optical amplifier can have a large but finite bandwidth. The size of the bandwidth depends on the material gain spectra, and hence the performance limit in terms of the bandwidth is one that can fully utilise the whole material gain spectrum. As mentioned in subsection 2.3.2, the structure of an amplifier must be designed properly so that the amplifier bandwidth matches that of the material gain. One such semiconductor laser diode amplifier structure is the so-called travelling-wave amplifier (TWA). The principle behind the TWA's capability to utilise the full possible bandwidth will be discussed in detail in subsequent chapters.

2.5.3 Performance limits of the saturation output power

The limit on the maximum saturation output power which can be achieved in real devices depends strongly on the gain saturation mechanisms [55]. It also depends on the maximum amplifier gain which can be obtained because, as we have seen, the gain–saturation mechanisms also determine the maximum gain that can be achieved by the device. Usually, a high-gain optical amplifier will have a high saturation output power because a high level of population inversion can be maintained for a wide range of input and output power in amplifiers [12]. However, there are also exceptions to this postulate (e.g. Fabry–Perot types of semiconductor laser diode amplifiers) [17, 24].

2.5.4 Performance limits of the noise figure

In contrast to the above performance limits of gain, bandwidth and saturation output power, which cannot reach the ideal infinite limits, the noise figure of an optical amplifier can reach the theoretical 0 dB (i.e. noiseless) value even though spontaneous emissions are present [6]. This rather surprising characteristic will be discussed later. However, at this stage it should be noted that, as shown in reference [6], most of the optical amplifiers cannot achieve this noise figure unless they satisfy certain conditions, such as having a phase-sensitive gain, as in parametric amplifiers [10]. When a semiconductor laser diode amplifier or a fibre amplifier is used in a linear application (i.e. as an in-line repeater), the minimum noise figure which can be achieved is 3 dB. The amplifiers that satisfy this minimum noise figure requirement are said to be operating in the quantum limit. This quantum limit and the corresponding noise figure can be derived by quantum mechanics [6]. However, in what follows we present a more intuitive approach to obtain this limiting value of the noise figure [43]. Consider the input electric field E_{in} of the light signal into an optical amplifier. This field can be expressed in terms of quadrature components as in [6, 42]

$$E_{in} = A_1 \cos(\omega t) + A_2 \sin(\omega t) = R \sin(\omega t + \theta) \qquad (2.29)$$

where

$$\omega = 2\pi f \qquad (2.30a)$$

$$R = \sqrt{A_1^2 + A_2^2} \qquad (2.30b)$$

$$\theta = \sin^{-1}\left[\frac{A_1}{\sqrt{A_1^2 + A_2^2}}\right] \qquad (2.30c)$$

At the output of the amplifier both quadrature components are amplified. If the gain for the cosine component is G_1 and that for the sine component is G_2 we can express the output electric field of the amplifier E_{out} as

$$E_{out} = (\sqrt{G_1}A_1 + \delta_{1i}) \cos(\omega t) + (\sqrt{G_2}A_2 + \delta_{2i}) \sin(\omega t) \qquad (2.31)$$

where δ_{1i} and δ_{2i} account for the amplitude and phase fluctuations due to the ith spontaneous emission event in the amplifier. Figures 2.5 and 2.6 illustrate the meaning of equation (2.31). In Figure 2.5 a phasor representation of E_{out} is shown, illustrating the effect of the random phase of the ith spontaneous emission ϕ_i on the amplitude and phase θ of the electric field of the signal [38, 51]. This fluctuation in amplitude and phase can be accounted for by the two random variables δ_{1i} and δ_{2i} for both quadrature components, as shown in Figure 2.6. Figure 2.6(a) represents an amplification process with $G_1 = G_2$ and when δ_{1i} and δ_{2i} are both zero. An amplifier with $G_1 = G_2$ is known as a phase-insensitive linear amplifier [6]. This relation is true for both semiconductor laser diode amplifiers and fibre amplifiers in most applications. It can be seen from Figure 2.6(a) that the phase of E_{out} is the same as E_{in} and $G = G_1 + G_2$. In

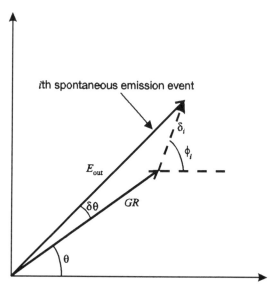

Figure 2.5 Phasor representation of the electric field in an optical amplifier contaminated by spontaneous emissions

Figure 2.6(b), the amplified field of a phase-insensitive amplifier with non-zero δ_{1i} and δ_{2i} is shown, where it can be seen that the phase of E_{out} has been altered by these spontaneous emissions. Notice the similarities between Figure 2.5 and Figure 2.6(b), where they actually represent an identical situation, but in different mathematical representations (phasor form in Figure 2.5 and quadrature components in Figure 2.6(b). Figure 2.6(c) shows amplification with phase-sensitive gain (e.g. in parametric amplification [6, 12]), where $G_1 \neq G_2$. Again, the field shows a fluctuation of amplitude and phase for the noiseless case. However, because of the phase sensitivity, G_1 and G_2 can be adjusted until the difference in amplification between the two phase quadrature components can counteract these fluctuations in amplitude and phase, resulting a 0 dB noise figure. By knowing the physical meaning of equation (2.31) we can proceed to analyse the noise figure of an amplifier as follows. Because δ_{1i} and δ_{2i} are random quantities, and as the signal-to-noise ratio at the amplifier output has to be obtained, we can measure the output power of the amplifier to see the overall effects of these random spontaneous emission events. This involves an ensemble average of E_{out}^2 which is given by [6]

$$\langle E_{\text{out}}^2 \rangle = \langle (\sqrt{G_1} A_1 + \delta_{1i})^2 \rangle + \langle (\sqrt{G_2} A_2 + \delta_{2i})^2 \rangle \qquad (2.32)$$

where $\langle \cdot \rangle$ represents a normalised ensemble average process, that is $\langle \cos^2(\omega t + \theta) \rangle = \langle \sin^2(\omega t + \theta) \rangle = 1$. Such an ensemble averaging includes a summation for all possible spontaneous emission events during the measuring period [38, 51]. Therefore, the ensemble average of the cross product of the sine and cosine quadrature terms goes to zero in the above equation. The above

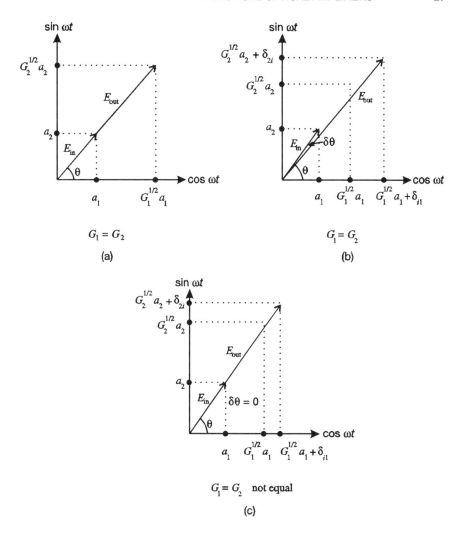

Figure 2.6 Quadrature representation of the electric field in (a) a phase-insensitive amplifier; (b) a phase-insensitive amplifier with fluctuations due to spontaneous emissions; and (c) a phase amplifier ($G_1 \neq G_2$) with fluctuations

process usually occurs when the output quadrature components are detected by coherent methods [6, 42].

Suppose that A_1 and A_2 in the input signals are contaminated with noise. This can be represented by two fluctuating parameters ΔA_1 and ΔA_2 for both phase quadratures, respectively [39]. Equations (6.29) and (6.31) can be re-written as

$$E_{in} = (A_1 + \Delta A_1) \cos(\omega t) + (A_2 + \Delta A_2) \sin(\omega t) \tag{2.33}$$

$$E_{\text{out}} = [\sqrt{G_1}(A_1 + \Delta A_1) + \delta_{1i}]\cos(\omega t) + [\sqrt{G_2}(A_2 + \Delta A_2) + \delta_{2i}]\sin(\omega t)$$

$$(2.34)$$

In determining the noise figure of the amplifier, the power at the input has to be determined. This can be deduced from equation (2.33) as

$$\langle E_{\text{in}}^2 \rangle = [A_1^2 + A_2^2] + [\langle \Delta A_1^2 \rangle + \langle \Delta A_2^2 \rangle] \tag{2.35}$$

In the above equation the first term in brackets is the input signal power and the second term in brackets is the input noise power. Also we have assumed that the noise and signal of the amplifier are uncorrelated such that $\langle A_1 \Delta A_1 \rangle = \langle A_2 \Delta A_2 \rangle = 0$ [6]. In addition, A_1 and A_2 are considered to be stationary quantities [59]. By using the same assumption about A_i and ΔA_i, as well as a similar assumption between A_i and δ_{ji}, and ΔA_i and δ_{ji} $(j = 1, 2)$, equation (2.34) can be simplified as

$$\langle E_{\text{out}}^2 \rangle = [G_1 A_1^2 + G_2 A_2^2] + [G_1 \langle \Delta A_1^2 \rangle + G_2 \langle \Delta A_2^2 \rangle] + [\langle \delta_{1i}^2 \rangle + \langle \delta_{2i}^2 \rangle] \tag{2.36}$$

In equation (2.36) the first term in brackets is the amplified output signal power, the second term in brackets is the amplified input noise power and the third term in brackets is the noise power due to spontaneous emissions. From equations (6.35) and (6.36) the signal-to-noise ratios at the input and output of the amplifier can be found, respectively, to be

$$(S/N)_{\text{in}} = \frac{(A_1^2 + A_2^2)}{(\Delta A_1^2 + \Delta A_2^2)} \tag{2.37}$$

and

$$(S/N)_{\text{out}} = \frac{G_1 A_1^2 + G_2 A_2^2}{G_1 \langle \Delta A_1^2 \rangle + G_2 \langle \Delta A_2^2 \rangle + \langle \delta_{1i}^2 \rangle + \langle \delta_{2i}^2 \rangle} \tag{2.38}$$

Consider a phase-insensitive amplifier with $G = G_1 = G_2$. Then the noise figure F is given by

$$F = \frac{(S/N)_{\text{in}}}{(S/N)_{\text{out}}} = \frac{G[\langle \Delta A_1^2 \rangle + \langle \Delta A_2^2 \rangle] + \langle \delta_{1i}^2 \rangle + \langle \delta_{2i}^2 \rangle}{G[\langle \Delta A_1^2 \rangle + \langle \Delta A_2^2 \rangle]} \tag{2.39}$$

When an amplifier is operating at the quantum limit (i.e. a single noise photon can be detected) the uncertainty principle [8] tells us that we cannot distinguish between a photon due to spontaneous emission and one due to the amplified input noise. Under these circumstances we can postulate that after ensemble averaging we have [6];

$$\langle \Delta A_1^2 \rangle = \frac{\langle \delta_{1i}^2 \rangle}{G} \tag{2.40a}$$

$$\langle \Delta A_2^2 \rangle = \frac{\langle \delta_{2i}^2 \rangle}{G} \tag{2.40b}$$

Substituting equation (2.40) into equation (2.39) gives $F = 2$ (i.e. 3 dB). Hence the basic cause of this limit of the noise figure of an optical amplifier is because of the amplification of uncertainty (or fluctuation) at the input and output [6]. For most optical amplifier applications this represents the minimum noise figure that any real device can ultimately achieve. *F* can reach 0 dB if the gain is phase sensitive (i.e. $G_1 \neq G_2$). This corresponds to optical amplifiers operating as parametric amplifiers [10, 60, 61], which is beyond the scope of the present book.

2.6 SUMMARY

In this chapter the most important properties of an optical amplifier have been discussed. They are (i) amplifier gain, (ii) bandwidth, (iii) saturation output power and (iv) noise figure. The particular importance of noise figure has been highlighted, and has been shown to be the parameter that is most significant in limiting the performance of any system incorporating optical amplifiers.

A two-level system has been used to discuss how these four properties arise in optical amplifiers. Although a more detailed treatment of optical amplifiers will be given in subsequent chapters, this simple model is sufficient to give a basic physical understanding of the fundamental principles of optical amplifiers.

The ideal performance characteristics of optical amplifiers have also been explored. To design real devices which can perform closely to this ideal, quantitative criteria are needed. These ideal characteristics are therefore taken further from qualitative terms to quantitative descriptions of the ultimate performance limits of an optical amplifier. It has been seen that the structure and material characteristics of the amplifier play an important role in determining its ultimate performance. Finally, the ultimate noise figure of an optical amplifier operating at the quantum limit has been analysed using a rather simplistic and intuitive approach. This quantum limit of optical amplifiers determines the physical limit at which a device can perform.

2.7 REFERENCES

[1] A. Yariv, *Quantum Electronics*, 3rd edition, John Wiley and Sons, New York, 1989.
[2] N. Nakagawa and S. Shimada, "Optical amplifiers in future optical communication systems", *IEEE LCS Mag.*, **1**, No. 4, pp. 57–62, 1990.
[3] C. H. Henry and B. H. Verbeck, "Solution of the scalar wave equation for arbitrarily shaped dielectric waveguides by two-dimensional Fourier analysis", *IEEE J. Lightwave Technol*, **LT-7**, No. 32, pp. 308–313, 1989.
[4] D. Marcuse, "Derivation of analytical expression for the bit-rate probability of lightwave systems with optical amplifiers", *IEEE J. Lightwave Technol.*, **LT-8**, No. 12, pp. 1816–1823, 1990.
[5] A. Yariv, "Signal-to-noise considerations in fiber links with periodic or distributed optical amplification", *Opt. Lett.*, **15**, No. 19, pp. 1064–1066, 1990.
[6] Y. Yamamoto and T. Mukai, "Fundamental of optical amplifiers", *Opt. Quantum Electron.*, **QE-21**, pp. S1–S14, 1989.
[7] M. I. Sargent, M. O. Scully and W. E. Lamb Jr, *Laser Physics*, Addison-Wesley, Wokingham, UK, 1974.

[8] W. H. Louisell, *Quantum Statistical Properties of Radiation*, John Wiley and Sons, New York, 1973.

[9] R. Loudon, *The Quantum Theory of Light*, 2nd edition, Oxford University Press, Oxford, UK, 1983.

[10] A. Yariv, *Optical Electronics*, 3rd edition, Holt–Saunders, 1985.

[11] D. Marcuse, "Computer model of an injection laser amplifier", *IEEE J. Quantum Electron.*, **QE-19**, No. 1, pp. 63–73, 1983.

[12] G. H. B. Thompson, *Physics of Semiconductor Laser Devices*, John Wiley and Sons, New York, 1980.

[13] Y. Mukai, Y. Yamamoto and T. Kimura, "Optical amplification by semiconductor lasers", *Semiconductors and Semimetals*, **22**, Part E, pp. 265–319, Academic Press, London, 1985.

[14] T. Mukai and Y. Yamamoto, "Gain, frequency bandwidth, and saturation output power of AlGaAs DH laser amplifiers", *IEEE J. Quantum Electron.*, **QE-17**, No. 6, pp. 1028–1034, 1981.

[15] T. Mukai and Y. Yamamoto, "Noise in an AlGaAs semiconductor laser amplifier", *IEEE J. Quantum Electron.*, **QE-18**, No. 4, pp. 564–575, 1982.

[16] G. Eisenstein, "Semiconductor optical amplifiers", *IEEE Circuits and Devices Mag.*, pp. 25–30, 1989.

[17] M. J. O'Mahony, "Semiconductor laser optical amplifiers for use in future fibre systems", *IEEE J. Lightwave Technol.*, **LT-6**, No. 4, pp. 531–544, 1988.

[18] J. L. Gimlett and N. K. Cheung, "Effects of phase-to-intensity noise conversion by multiple reflections on giga-bit-per-second DFB laser transmission system", *J. Lightwave Technol.* **LT-7**, No. 6, pp. 888–895, 1989.

[19] J. L. Gimlett, M. Z. Iqbal, L. Curtis, N. K. Cheung, A. Fontana and G. Grasso, "Impact of multiple reflection noise in Gbit/s lightwave system with optical fibre amplifiers", *Electron. Lett.*, **25**, No. 20, pp. 1393–1394, 1989.

[20] A. Elrefaire and C. Lin, "Performance degradations of multi gigabit-per-second NRZ/RZ lightwave system due to gain saturation in travelling-wave semiconductor optical amplifiers", *IEEE Photon. Technol. Lett.*, **1**, No. 10, pp. 300–303, 1989.

[21] M. J. Adams, "Time dependent analysis of active and passive optical bistability in semiconductors", *IEE Proc.*, **132**, Part J, No. 6, pp. 343–348, 1985.

[22] J. W. Crowe and R. M. Craig, "Small-signal amplification in GaAs lasers", *Appl. Phys. Lett.*, **4**, No. 3, pp. 57–58, 1964.

[23] W. J. Crowe and W. E. Aheam, "Semiconductor laser amplifier", *IEEE J. Quantum Electron.*, **QE-2**, No. 8, pp. 283–289, 1966.

[24] J. Wang, H. Olesen and K. E. Stubkjaer, "Recombination, gain and bandwidth characteristics of $1.3\,\mu m$ semiconductor laser amplifiers", *IEEE J. Lightwave Technol.*, **LT-5**, No. 1, pp. 184–189, 1987.

[25] T. Mukai, Y. Yamamoto and T. Kimura, "S/N and error rate performance in AlGaAs semiconductor laser preamplifier and linear repeater systems", *IEEE Trans. Microwave Theory Technol.*, **MTT-30**, No. 10, pp. 1548–1556, 1982.

[26] H. Ghafouri-Shiraz and C. Y. J. Chu, "Analysis of waveguiding properties of travelling-wave semiconductor laser amplifiers using perturbation technique", *Fiber and Integrated Optics*, **11**, pp. 51–70, 1992.

[27] D. Marcuse, *Theory of Dielectric Optical Waveguides*, Academic Press, London, 1974.

[28] M. J. Adams, *An Introduction to Optical Waveguides*, John Wiley and Sons, New York, 1981.

[29] C. J. Hwang, "Properties of spontaneous and stimulated emissions in GaAs junction laser; I & II", *Phys. Rev. B*, **2**, No. 10, pp. 4117–4134, 1970.

[30] F. Stern, "Calculated spectral dependence of gain in excited GaAs", *J. Appl. Phys.*, **47**, No. 2, pp. 5382–5386, 1976.

[31] M. Osinski and M. J. Adam, "Gain spectra of quaternary semiconductors", *IEE Proc.* **129**, Part J, No. 6, pp. 229–236, 1982.

[32] I. D. Henning, M. J. Adams and J. V. Collins, "Performance prediction from a new amplifier model", *IEEE J. Quantum Electron.*, **QE-21**, No. 6, pp. 609–613, 1985.

[33] T. Saitoh and M. Mukai, "1.5 μm GaInAs travelling-wave semiconductor laser amplifier", *IEEE J. Quantum Electron.*, **QE-23**, No. 6, pp. 1010–1020, 1987.

[34] T. Mukai, K. Inoue and T. Saitoh, "Homogeneous gain saturation in 1.5 μm InGaAsP travelling-wave semiconductor laser amplifiers", *Appl. Phys. Lett.*, **51**, No. 6, pp. 381–383, 1987.

[35] R. Frankenberger and R. Schimpe, "Origin of non-linear gain saturation in index-guided InGaAsP laser diodes", *Appl. Phys. Lett.*, **60**, No. 22, pp. 2720–2722, 1992.

[36] G. Lasher and F. Stern, "Spontaneous and stimulated recombination radiation in semiconductors", *Phys. Rev.*, **133**, No. 2A, pp. 553–563, 1964.

[37] T. Mukai and T. Yamamoto, "Noise characteristics of semiconductor laser amplifiers", *Electron. Lett.*, **17**, No. 1, pp. 31–33, 1981.

[38] C. H. Henry, "Theory of the phase noise and power spectrum of a single mode injection laser", *IEEE J. Quantum Electron.*, **QE-19**, No. 9, pp. 1391–1397, 1983.

[39] K. Hinton, "Optical carrier linewidth broadening in a travelling wave semiconductor laser amplifier", *IEEE J. Quantum Electron.*, **QE-26**, No. 7, pp. 1176–1182, 1990.

[40] K. Kikuchi, C. E. Zah and T. P. Lee, "Measurement and analysis of phase noise generated from semiconductor optical amplifiers", *IEEE J. Quantum Electron.*, **QE-27**, No. 3, pp. 416–422, 1991.

[41] A. J. Lowery, "Amplified spontaneous emission in semiconductor laser amplifiers: validity of the transmission-line laser model", *IEE Proc.*, **137**, Part J, No. 4, pp. 241–247, 1990.

[42] H. Taub and D. L. Schilling, "Principles of Communication Systems", 2nd edition, McGraw-Hill, New York, 1986.

[43] C. Y. J. Chu, "Semiconductor laser optical amplifiers for optical communications", M.Phil. (Eng.), qualifying thesis, University of Birmingham, UK, 1991.

[44] Y. Yamamoto, "Noise and error rate performance of semiconductor laser amplifiers in PCM-IM optical transmission systems", *IEEE J. Quantum Electron.*, **QE-16**, No. 10, pp. 1073–1081, 1980.

[45] D. Marcuse, "Calculation of bit-error probability for a lightwave system with optical amplifiers", *IEEE J. Lightwave Technol.*, **9**, No. 4, pp. 505–513, 1991.

[46] A. Lord and W. A. Stallard, "A laser amplifier model for system optimization", *Opt. Quantum Electron.*, **QE-21**, pp. 463–470, 1989.

[47] O. Lumholt, K. Schusler, A. Bjarklev, S. Dahl-Petersen, J. H. Poulsen, T. Rasmussen and K. Rottwitt, "Optimum position of isolators within erbium-doped fibres", *IEEE Photon. Technol. Lett.*, **4**, No. 6, pp. 568–569, 1992.

[48] D. J. Maylon and W. A Stallard, "565 Mbit/s FSK direct detection system operating with four cascaded photonic amplifiers", *Electron. Lett.*, **25**, No. 8, pp. 495–497, 1989.

[49] K. Vahala and A. Yariv, "Semiclassical theory of noise in semiconductor lasers-Part I and II", *IEEE J. Quantum Electron.*, **QE-19**, No. 6, pp. 1096–1109, 1983.

[50] P. Spano, S. Piazzola and M. Tamburrini, "Phase noise in semiconductor lasers; a theoretical approach", *IEEE J. Quantum Electron.*, **QE-19**, No. 7, pp. 1195–1199. 1983.

[51] C. H. Henry, "Phase noise in semiconductor lasers", *IEEE J. Lightwave Technol.*, **LT-4**, No. 3, pp. 298–311, 1986.

[52] B. Glance, G. Eisenstein, P. J. Fitzgerald, K. J. Pollack and G. Raybon, "Optical amplification in a multichannel FSK coherent system", *Electron. Lett.*, **24**, No. 18, pp. 1157–1159, 1988.

[53] M. A. Ali, A. F. Elrefaie and S. A. Ahmed, "Simulation of 12.5 Gb/s lightwave optical time-division multiplexing using semiconductor optical amplifiers as external modulators", *IEEE Photon. Technol. Lett.*, **4**, No. 3, pp. 280–282, 1992.

[54] R. M. Fortenberry, A. J Lowery and R. S. Tucker, "Up to 16 dB improvement in detected voltage using two section semiconductor optical amplifier detector", *Electron. Lett.*, **28**, No. 5, pp. 474–476, 1992.

[55] G. P. Agrawal, "Effect of gain and index nonlinearities on single-mode dynamics in semiconductor lasers", *IEEE J. Quantum Electron.*, **QE-26**, No. 11, pp. 1901–1909, 1990.

[56] M. J. Adams, J. V. Collins and I. D. Henning, "Analysis of semiconductor laser optical amplifiers", *IEE Proc.*, **132**, Part J, No. 1, pp. 58–63, 1985.

[57] A. J. Lowery, "Modelling spectral effects of dynamic saturation in semiconductor laser amplifiers using the transmission-line laser model", *IEE Proc.*, **136**, Part J, No. 6, pp. 320–324, 1989.

[58] Z. Pan, H. Lin and M. Dagenais, "Switching power dependence on detuning and current in bistable diode laser amplifiers", *Appl. Phys. Lett*, **58**, No. 7, pp. 687–689, 1991.

[59] D. L. Snyder, *Random Point Processes*, John Wiley and Sons, New York, 1975.

[60] Y. Yamamoto (Editor), *Coherence, Amplification, and Quantum Effects in Semiconductor Lasers*, John Wiley and Sons, New York, 1991.

[61] K. Kikuchi, "Proposal and performance analysis of novel optical homodyne receiver having an optical preamplifier for achieving the receiver sensitivity beyond the shot-noise limit", *IEEE Photon. Technol. Lett.*, **4**, No. 2, pp. 195–197, 1992.

OPTICAL AMPLIFICATION IN SEMICONDUCTOR LASER DIODES

3.1 INTRODUCTION

In Chapter 2 the major characteristics of optical amplifiers were explored. These characteristics are important in determining the ultimate performance of any optical communication system using optical amplifiers [1]. However, as discussed in the previous chapter, the exact performance requirements of an optical amplifier vary with its type and the application for which it is used. In this chapter we continue our discussion by considering how optical amplification can be achieved in semiconductor lasers. This will help us to understand the physical processes that determine the characteristics of semiconductor laser amplifiers (SLAs), and hence the performance requirements in different SLA applications. The mechanisms of optical gain in semiconductor materials will first be reviewed, and we will discuss how optical amplification can be realised in semiconductor laser structures. A summary of the historical development of SLAs from the literature will then be reviewed, and different types of semiconductor laser amplifiers will also be examined. Finally, a review of current research trends in different applications, and the subsequent performance requirements of SLAs, will be presented.

3.2 PRINCIPLES OF OPTICAL AMPLIFICATION IN SEMICONDUCTOR LASERS

It is well known in linear systems theory that an oscillator is essentially an amplifier with positive feedback. The same is true for laser diode oscillators. The gain of an amplifier with positive feedback, as mentioned in Chapter 2, is given by

$$G = \frac{G_s}{1 + FG_s} \tag{3.1}$$

where G_s is the forward path gain and F is the amount of feedback gain. Equation (3.1) indicates that oscillation will occur when $FG_s = -1$. As long as $|FG_s| < 1$, the amplifier will not oscillate and will provide gain according to equation (3.1). This is true for semiconductor lasers as well. If one can reduce F, and/or G_s, such that the laser is operating below the oscillation threshold, then the laser can be used as a simple optical amplifier for light signals [2]. Historically, amplification using semiconductor lasers stemmed from this idea of using the laser below the oscillation threshold. Therefore the early development of SLAs was closely related with that of semiconductor laser diodes (SLDs), and in many early works a SLA is identical to a SLD in terms of their structures. If we are to understand more fully the principles of optical amplification using semiconductor lasers, we must first understand (i) how optical gain is achieved in semiconductors and (ii) the structural aspects of semiconductor lasers that will affect *both* the values of F and G_s. We now review these two aspects briefly to provide an introduction to the results of further investigations which will be presented later.

3.2.1 *Optical processes in semiconductors*

As we have seen in Chapter 2, population inversion is the vital condition to provide optical amplification. In semiconductors, population inversion is formed by forward biasing a heavily doped p–n junction [3, 4], the band diagram of which is shown in Figure 3.1. For such a heavily doped *degenerate* semiconductor, the Fermi energy level lies above the conduction band edge in an n-type material, and below the valence band edge in a p-type material [5]. When these two types of material are joined to form a p–n junction, the diffusion of minority carriers occurs (electrons toward the p-side, and holes toward the n-side). The diffusion stops when the resulting build-up of an electric field can counteract the diffusion process (Figure 3.1(a)). This field is known as the *depletion* electric field, and can be lowered by applying a forward bias across the junction as shown in Figure 3.1(b). In addition to the diffusion process, a drift of minority carriers across the junction occurs. Because of the positions of the Fermi energy levels, these carrier movements will create a very narrow region in which there are both electrons in the higher energy conduction band and holes in the valence band, a situation that is impossible for an ordinary p–n junction. In effect, a population inversion occurs in this region. Consequently this region is known as the *active* region where optical amplification becomes possible. This structure is referred to as a *homojunction* because the bandgap energy E_g on both the p and n sides of the structure is the same [4].

Because the population inversion level depends on the number of injected carriers brought by the forward bias, the corresponding optical gain is directly related to the injected carrier density, which is usually denoted by n (measured

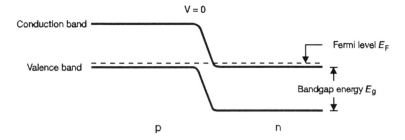

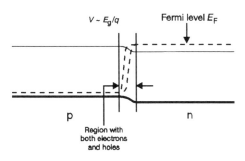

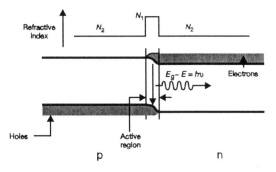

Figure 3.1 (a) Unbiased p–n junction under thermal equilibrium. (b) Forward biased heavily doped p–n junction. (c) Formation of an active region under forward bias.

in cm^{-3}). Moreover, population inversion is seldom complete because of processes like non-radiative recombinations and diffusion [4, 5]. The degree of inversion is described by a parameter known as the *population inversion parameter* n_{sp}, which is defined as

$$n_{sp} = \frac{n}{n - n_0} \tag{3.2}$$

where n_0 is known as the *transparency carrier density*. This parameter can also describe the amount of induced absorption taking place. When the injected carrier density n is less than n_0, the rate of absorption of photons is higher than that of stimulated emissions and no amplification of light is possible. The above parameter therefore measures the degradation of the available carrier density from the injected value due to absorption processes [6].

The refractive index of the region which provides population inversion is also higher than that of the surrounding layers because of the excess electrons, or carrier density, as predicted by the Kramers–Kronig relation [7]. Hence a refractive index step exists between the active region and the surroundings (Figure 3.1(c)), which effectively forms a dielectric waveguide to confine the light propagating along the structure [5]. Consider an optical signal injected in the plane perpendicular to the p–n junction into the SLA. Two physical processes occur. The refractive index difference will guide the signal across the active region, acting as an optical waveguide [8]. As the signal travels across the active region population inversion will generate coherent photons by stimulated emissions and amplify the signal by a gain G_s. Because of the presence of such a dielectric optical waveguide within the structure, different modes can propagate across the structure, each with a different group velocity [8–10]. Modern laser structures can minimise the number of modes that can propagate to two by tailoring the active region dimensions [4, 5, 11]. It can be shown that the electric fields of these modes are orthogonal to each other, and we regard each of these modes as having different *states of polarisation* [12]. We will discuss this in further detail in the following chapter. It should be noted, however, that because the active region can at least support the propagation of signals with two different states of polarisation having different group velocities (hence confinement within the structure), we might also expect the gain for these two states of polarisation to be different. The resulting *polarisation sensitivity* of the optical gain in a laser amplifier is one of the crucial areas that has been addressed recently by many research workers [13–16], and will be discussed in depth in Chapter 4.

In addition to losses due to induced absorption, photons can be lost via scattering [4]. The overall gain in the active region can be quantified by two material parameters: the *material gain coefficient* g_m and *loss coefficient* α. Both quantities are measured in cm^{-1}, and for a distance L, these two parameters are related to G_s by the following relation:

$$G_s = \exp(g_m - \alpha)L \qquad (3.3)$$

In the above equation we have assumed a perfect guiding condition, i.e. all signals are confined to the active region by the waveguiding action [8]. In practice, because the waveguide formed is only a weakly guiding one, with $(N_1 - N_2) \ll N_1$ (see Figure 3.1(c)), power will "leak" out of the active region into the cladding (the bulk of the dielectric material surrounding the active region) due to evanescent fields [9]. To take this into account, equation (3.3) is modified by a weighting factor Γ such that

$$G_s = \exp[\Gamma(g_m - \alpha_a) - (1 - \Gamma)\alpha_c]L \qquad (3.4)$$

where α_a and α_c are the absorption coefficients in the active and claddings regions, respectively. The factor Γ is known as the optical confinement factor, and is defined as the ratio between the optical power confined in the active region to the total optical power flowing across the structure [8, 17]. The form of equation (3.4) suggests that the proportion of power in the active region is related to the magnitude of the material gain coefficient of the structure, and an alternative expression for more complex structures has been derived by the author using perturbation theory [18]. Alternatively, it can be shown that equation (3.4) can be directly obtained by analysing the power flow using Poynting vectors [8]. Furthermore, as we have discussed previously, the state of polarisation of the signals will affect the optical gain of the amplifier. In practice, it is the optical confinement factor Γ that is actually affected by the polarisation of the signals. Hence the value of G_s is polarisation-dependent according to equation (3.4). This is particularly important when the facet reflectivities of the amplifiers are zero (see subsection 3.4.1), in which case the overall gain of the amplifier is $G = G_s$.

3.2.2   Analysis of the optical gain in semiconductors

The exact mathematical formulation of the material gain coefficients in semiconductors has been investigated since the early developments in semiconductor lasers and laser amplifiers [19–21]. In practice, population inversion in semiconductor lasers cannot be described quantitatively in a straightforward manner. Because of the presence of bands rather than sharp and distinct energy levels, the density of states and hence the population densities of carriers in both the conduction and valence bands are spread non-uniformly over a range of photon energies [5]. This statistical spread of the carrier density results in different stimulated emission rates for different incident photon energies (hence signal frequencies and wavelengths). This spectral function of stimulated emission will ultimately determine the material gain coefficient, and has to be found by Fermi–Dirac statistics with some assumptions on the band structure as well as the nature of the radiative transitions [4, 22]. A detailed formulation is clearly beyond the scope of the present book, though the interested reader can find a brief description in section 3.6. In the following discussion we will describe some of the most useful approximations which can be used conveniently in analysing the material gain coefficients in semiconductor laser devices without using complex statistical and quantum mechanical methods.

The results of the exact calculation [21] of the *peak* material gain coefficient g_{peak} can be approximated by the following expression [3, 23]:

$$g_{peak} = A(n - n_0) \tag{3.5}$$

where A is a constant. This linear relationship gives an accurate description of the relation between the material gain and the injected carrier density except in the range close to the transparency region, i.e. when g is very small. To overcome this problem a more accurate parabolic approximation of the form

$$g_{peak} = an^2 + bn + c \tag{3.6}$$

Table 3.1 Semiconductor laser parameters used to calculate Figure 3.2

Wavelength	λ	1.3 μm
Temperature	T	300 K
Coefficients		
A		7.2543×10^{-16} cm^{-2}
n_0		1.5034×10^{18} cm^{-3}
a		3.9138×10^{-34} cm^5
b		-8.2139×10^{-18} cm^2
c		428.7016 cm

where a, b and c are constants, has been suggested [24]. The constants in equation (3.6) can be found by a least-squares fitting technique to the available exact solution [19]. We have compared the difference between using equation (3.5) and equation (3.6) by calculating the peak gain coefficients of a SLA with the parameters listed in Table 3.1 [25], and the results are plotted in Figure 3.2. Also shown are the exact solutions calculated in reference [19]. As shown in Figure 3.2, a better fit can be obtained for a wider range of values of g_{peak} and n using equation (3.6), especially for small values of n.

In the absence of a significant degree of stimulated emission (below the threshold in a semiconductor laser, or no signal input if the laser is used as an amplifier), the injected carrier density is related to the injection or bias current i by

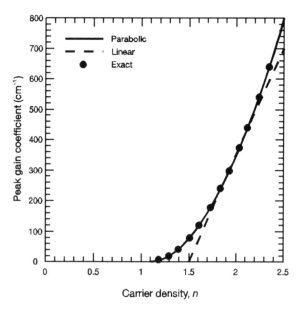

Figure 3.2 Comparison between linear and parabolic models for the peak gain coefficient calculated using the parameters listed in Table 3.1

$$i = \frac{eVR}{\eta_i} \tag{3.7}$$

where e is the electron charge, η_i is the internal quantum efficiency of the material [5], V is the volume of the active region and R is the *carrier recombination rate* expressed as [4]

$$R = \frac{n}{\tau_n} \tag{3.8}$$

where τ_n is the recombination lifetime of carriers. Using equations (3.7) and (3.8), the peak material gain coefficient can be calculated for a particular value of the injection current i for a particular material (which determines the values of η_i and τ_n) and structure (which determines the value of V).

According to the simple two-level model discussed in Chapter 2, carrier recombinations are due purely to radiative spontaneous emissions in the active region. If this is true, then according to the analysis outlined in section 3.5 [5]

$$R = Bnp \tag{3.9}$$

where B is known as the *radiative recombination coefficient* [26]. For an undoped semiconductor, $n = p$ and the spontaneous recombination rate is given by Bn^2. Hence B *is also known as* the bi-molecular *recombination coefficient* [27]. However, such a simple derivation ignores the effect of non-radiative recombinations due to, for example, the presence of recombination sites or defects in the material. In SLAs made of AlGaAs/GaAs (working at a wavelength of $0.85\,\mu m$), these non-radiative recombinations of carriers are unimportant. However, for InGaAsP/InP SLAs operating at a longer wavelength of $1.3\,\mu m$ and $1.55\,\mu m$, the contribution of these non-radiative recombinations is quite significant.

There are two major sources of non-radiative recombinations. We have mentioned recombination with defects above. A far more important recombination process is *Auger recombination* [4], which involves four particle states (three electrons and one hole, or two electrons and two holes, etc). In this process the energy released during the electron–hole recombination is transferred to another electron or hole, which gets excited and then relaxes back to the lower state by losing its energy to lattice vibration or *phonons*. Both of these recombination processes are similar to spontaneous emissions in that they are all random events. However, in Auger recombinations and recombinations with defects, no photons are emitted. If these non-radiative recombinations are included, the recombination rate R can then be expressed by [4, 26, 27, 33]

$$R = A_{nr} n + Bn^2 + Cn^3 \tag{3.10}$$

The three terms represent recombination with defects or surface, spontaneous emission and Auger recombinations, respectively. The difference between using equation (3.9) and equation (3.10) on calculating the carrier recombination lifetime τ_n from equations (3.7) and (3.8) is illustrated by the curves in Figure 3.3. Here, the values of $1/\tau_n^2$ calculated by the two different models of

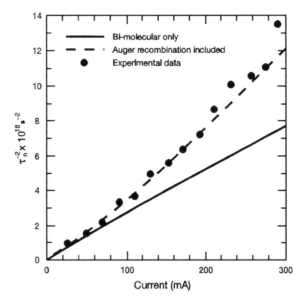

Figure 3.3 Comparison of the recombination lifetime calculated by the bi-molecular term only and by including non-radiative recombinations with experimental measurements obtained by Olshansky *et al.* [27]

equations (3.9) and (3.10) are plotted against the injection current to the semiconductor laser, where the parameters used are shown in Table 3.2, together with the values obtained from experiment [20]. The results in Figure 3.3 show that the bi-molecular model of equation (3.9) will underestimate the value of τ_n and hence overestimate the value of R for a particular value of injection current i. Wang *et al.* [26] concluded that equation (3.10) gives a better prediction of the performance of long wavelength semiconductor laser amplifiers, with better match to the experimental observations. Moreover, a more accurate value of R will also give a more accurate prediction of the actual material gain coefficient for a particular current i.

The material gain coefficient is affected by other factors in addition to carrier density. Because of the statistical distribution of the density of states, the exact

Table 3.2 Parameters of semiconductor laser used for comparison between different carrier recombination models

Laser length	200 μm
Active region width	25 μm
Active region thickness	0.2 μm
Wavelength	1.3 μm
Recombination coefficients	
A_{nr}	$2 \times 10^8 \, \text{s}^{-1}$
B	$0.38 \times 10^{-16} \, \text{m}^3 \, \text{s}^{-1}$
C	$0.51 \times 10^{-41} \, \text{m}^6 \, \text{s}^{-1}$

value of the material gain coefficient will depend on the optical frequency f (and hence wavelength λ) of the incident signal. So far, we have assumed that the gain is measured at a signal wavelength identical to that of the peak in the gain spectrum λ_0. Strictly speaking, the exact value of g differs from g_{peak} if λ is different from λ_0. A closed-form expression for the spectral function of the material gain coefficient in quaternary semiconductors is given by [28]

$$g_m(E) = \frac{\dfrac{\pi^3 c^2 (h/2\pi)^3}{8 N_g^2} C \left(\dfrac{\Delta E}{E}\right)^2 \sinh[(F_c + F_v - \Delta E)/2kT]}{\cosh[(F_c + F_v - \Delta E)/2kT] + \cosh[(F_c - F_v)/2kT]} \qquad (3.11)$$

where N_g is the group refractive index of the active region, F_c and F_v are the quasi-Fermi levels in the conduction and valence bands, respectively, ΔE is the energy difference between the signal photon and the band gap energy E_g (the energy difference between the conduction and valence band edges), T is the temperature, k is the Boltzmann constant and C is a constant that weights out the contribution of masses of electrons and holes in the spectral function of gain. This expression is true for semiconductors with parabolic bands and with no k-selection for radiative transitions [5, 7].

We have plotted the material gain spectrum obtained from equation (3.11) in Figure 3.4 for undoped $In_{1-x}Ga_x As_y P_{1-y}$ having $x = 0.25$ and $y = 0.55$ at room temperature ($T = 300$ K) for six different values of the carrier density n and two different wavelengths. It can be seen that the whole spectrum shifts upwards as n increases. There are two values of λ where g_m crosses over zero. First, there is a gradual decrease in the material gain with longer wavelength. Note that this cut-off point for the longer wavelength end is almost identical for all values of the injection carrier density (see Figure 3.4). This can be explained as follows. As the wavelength increases, the energy of the photon reduces until it is less than the bandgap energy of the material (Figure 3.1(c)). Emissions (and absorption) of photons with energies less than the bandgap energy are not allowed, and hence the optical gain reduces to zero for this particular value of the photon energy. This explains why, for all levels of injection, the cut-offs at the longer wavelength end are almost identical. The slight differences between these zero-crossings for different injection carrier densities arise because of the slight modification of the shape of the band edge (known as *band tailing*) with injection current [22].

The cut-off of $g_m = 0$ at the shorter wavelength end can be understood from the fact that the optical gain can only be maintained if the signal photon frequency f satisfies the basic constraint of (see section 3.6) [4, 7]

$$E = hf \leqslant (F_c - F_v) \qquad (3.12)$$

The upper limit on the frequency f in the above equation accounts for this shorter wavelength cut-off of the spectrum. As the carrier densities decrease for energy levels which are farther away from the band edge in both the conduction and valence bands, the stimulated emission rate also reduces for transitions involving larger photon energies. Hence the signal gain reduces with reducing wavelength until $g_m < 0$, beyond which absorption predominates over

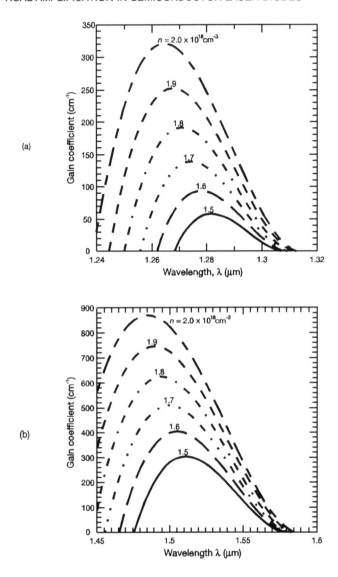

Figure 3.4 Gain spectra for undoped InGaAsP at $T = 300$ K using equation (3.11) for an injection carrier density ranging from 1.5×10^{18} to 2.5×10^{18} cm^{-3} for (a) 1.3 μm and (b) 1.55 μm wavelengths

stimulated emissions. Equation (3.11) can be approximated by the following parabolic form around the peak gain region [29]:

$$g_m(n, \lambda) = g_{peak}(n) - \gamma(\lambda - \lambda_0)^2 \tag{3.13}$$

where γ is a constant related to the actual spectral width of the material gain spectrum. An example of such fitting to one of the curves in Figure 3.4 is shown

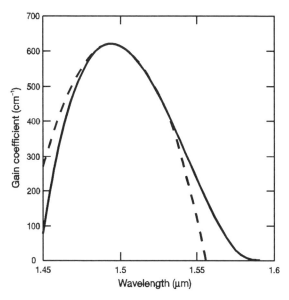

Figure 3.5 An example of the parabolic fit to an exact gain spectrum of undoped InGaAsP. The solid line represents gain coefficients calculated by equation (3.11) with $n = 1.8 \times 10^{18}$ cm^{-3} for the emission wavelength of 1.5 μm, whilst the dotted line represents the parabolic fit around the gain peak region

in Figure 3.5. Notice that from equation (3.13) we can see that g_m will be increased by the same proportion for all wavelengths in the gain spectrum as n increases. The gain spectrum is said to be *homogeneously broadened*. By increasing the injection current the peak gain value will increase until a point where saturation will occur (notice the difference between approaching saturation in this way with that discussed in Chapter 2). At this point the peak material gain coefficient cannot be increased further by increasing pumping, and is effectively clamped [7]. A common situation to observe this phenomenon is when lasing occurs in semiconductor lasers. It is found that in semiconductor lasers, when the peak gain coefficient has reached the saturation value, the gain coefficients for other optical wavelengths across the whole gain spectrum are also clamped, although they are still far from the saturation value. This phenomenon is known as *homogeneous gain saturation*. This property was also shown to be true in SLAs [30].

As shown by the exact solution of the gain spectrum in reference [45] and from the analytical approximation given by equation (3.11), g_{peak} is also a function of temperature T, since it will affect the band gap separation [103]. In addition, λ_0 is affected by the carrier density n, as observed from Figure 3.4. A more complete expression for the material gain can be found to be of the form [27]

$$g_m(\lambda, n) = [a(T)n^2 + b(T)n + c(T)] - \gamma[\lambda - \lambda_0(n)]^2 \qquad (3.14)$$

where we have used equation (3.6). Usually, we operate the SLA at a fixed temperature using electro-thermal control, and any change in g_m due to a change in T can be ignored. However, the wavelength effect on g_m cannot be ignored, especially for WDM applications. Henning *et al.* [29] proposed a linear relation for the peak gain wavelength with n:

$$\lambda_0(n) = \lambda_{0p} - \xi\left(1 + \frac{n}{n_{th}}\right) \tag{3.15}$$

where ξ is a constant, λ_{0p} is the wavelength of transparency, i.e. the value of λ_0 when $g_{peak} = 0$, and n_{th} is the lasing threshold carrier density.

3.3 SEMICONDUCTOR LASER DIODES AS OPTICAL AMPLIFIERS

The possibility of introducing optical gain into semiconductors led to the birth of lasers fabricated by homojunctions, as described in the previous section, in the early 1960s. At the same time it was realised that the active region in these devices can also be used as an active waveguide to provide optical amplification below the lasing threshold. This eventually gave birth to the first semiconductor laser amplifiers. In what follows we will briefly trace the development of semiconductor laser amplifiers, where we will find, not surprisingly, that their development was closely linked to advances in technologies related to semiconductor laser diodes.

3.3.1 Optical amplification using homojunctions

After the first successful operation of semiconductor laser diodes [31, 32] the development of laser diode amplifiers received considerable attention. In the early stages, SLDs were fabricated with homojunctions using GaAs as the amplifying material (see Figure 3.6) [4, 5]. These homojunctions required a large threshold current to achieve laser oscillation, and can only be used in continuous-wave (cw) operation under very low sub-zero temperatures. At the

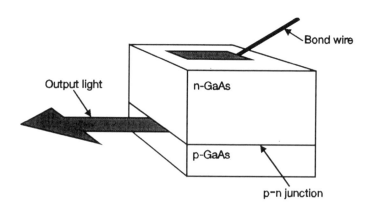

Figure 3.6 A homojunction semiconductor laser

same time, there were several successful attempts to use these homojunction SLDs as optical amplifiers by reducing the injection current, and hence the single pass gain G_s, below the threshold [6]. Crowe and Craig [31] reported a small signal amplification with GaAs laser diodes at 77 K. The injection current required to create the active region was found to be in the range of several amperes. A maximum small signal gain of $1000 (=30 \, \text{dB})$ can be obtained with an injection current density of $6000 \, \text{A/cm}^2$. The bandwidth of the amplifier was found to be 15 Å. An amplifier with a similar structure but higher gain was reported two years later [32]. The maximum signal gain reported at 77 K was 2000 with an output power of 150 mW. The bandwidth of the amplifier was 30 Å, twice that in the former experiment. In both experiments gain saturation occurred for high input light power with constant pumping (i.e. injection current). The gain–current characteristics were found to be exponential in both experiments, agreeing with the available theory at that time [32]. It was found that the gain measurement was very sensitive to coupling between the source and amplifier [31]. Side modes were observed at the output of the amplifier with a single mode input in the earlier experiment. Crowe and Ahearn [32] proposed the use of anti-reflection (AR) coatings by deposition of multi-layer dielectric three-quarter-wavelength coatings using SiO_2 to suppress these side modes. Thus the application of AR coatings in SLAs can be traced back to as early as 1966. They also discovered a shift in output wavelength of the amplifier towards the blue end of the optical spectrum as the pumping increases, which remained unexplained at that time.

Kosonocky and Cornely [33] reported a different device configuration as an optical amplifier. Their objective was to design an amplifier directly coupled to an oscillator. This was the first attempt to design a cleaved substrate amplifier [34]. A schematic of their device is shown in Figure 3.7. Again, their

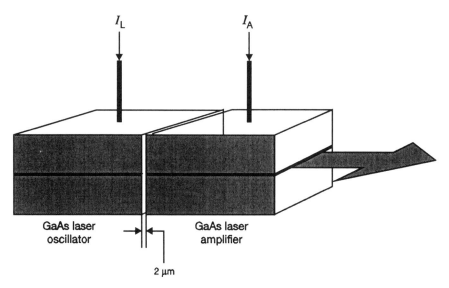

Figure 3.7   A cleaved-substrate semiconductor laser amplifier

experimental results showed a strong dependence of signal gain on input signal power. The maximum amplifier gain obtained was about 150 with an input signal power flux of $8\,\text{kW}/\text{cm}^2$. They also observed that spontaneous emission power was quenched as the input signal power increased, and the internal oscillation modes of the amplifier were suppressed as well. The amplifier gain was found to be inversely proportional to the separation between the oscillator and amplifier, since the coupling efficiency between the oscillator and the amplifier reduces with increasing separation. Their amplifier possessed a strong non-linear characteristic because of the non-uniform gain distribution along the longitudinal direction of the amplifier (see Chapter 4). Hence they proposed a technique for maintaining a uniform gain by exciting oscillations along the *transverse* section of the amplifier. Although this technique is not used now, their effort highlighted the problem faced by research workers in the 1960s, namely how to improve the gain uniformity along the longitudinal direction of the laser amplifier.

At the same time, much theoretical work was published introducing new ideas on the applications of optical amplifiers. For example, Arnaud [35] proposed that optical amplifiers could be used as pre-amplifiers to optical receivers to enhance receiver sensitivities. Personick [36] showed that by using optical amplifiers as in-line repeaters the performance of simple digital optical transmission using on–off keying could be improved. In addition, optical amplifiers could be used as front-end amplifiers for regenerative repeaters to improve their sensitivities. This early theoretical work laid down fundamentals for later work in system applications of optical amplifiers.

3.3.2   *Optical amplification using heterostructures*

The homojunction SLDs were soon found to be unsatisfactory, both as amplifiers and as oscillators, because of their operational requirements of low temperature, large threshold currents and difficulties in maintaining cw operation [5]. Both Kressel *et al.* in the United States and Alferov *et al.* in the U.S.S.R. discovered independently that the carrier recombinations, and hence threshold current, could be reduced greatly by introducing additional potential barriers around the active region [5]. Such structures proved to be a milestone in the development of modern semiconductor lasers and these are known as *heterojunctions* [4]. The most useful heterojunctions involve two potential barriers, one each on the p- and n-side, respectively, and are hence called *double heterostructures* (DH) (Figure 3.8). They were found to be extremely successful in avoiding excess carrier recombination, as well as providing waveguiding for the optical signals to travel across the active region, thereby improving the gain uniformity along the longitudinal direction [5]. The device can then be used as a cw source at room temperature. Although complicating the band structure, the basic principles behind the formation of and amplification in an active region remain unchanged. The equations described in previous sections can be applied to heterostructures without any significant modifications because the optical

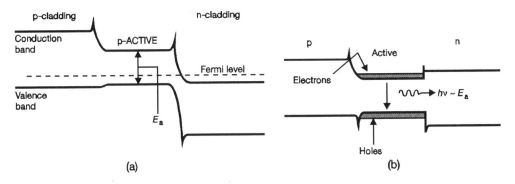

Figure 3.8 Band diagram for a double heterostructure at (a) zero bias and (b) forward bias

confinement factor Γ will take into account the effect of the structural effects on the gain of the amplifier [12].

Not surprisingly, experiments in optical amplification using DH semiconductor lasers biased below the oscillation threshold were carried out soon after successful operations of DH SLDs. For example, Schicketanz and Zeidler [37] developed a GaAs DH SLA which can operate under both travelling-wave (i.e. non-resonant; see subsection 3.4.1) and resonant conditions at room temperature. The non-resonant operations were achieved by immersing the amplifier in a fluid with a refractive index matched to the amplifier. They measured the amplified spontaneous emissions (ASE) at the amplifier output and found that the amplifier gain spectrum was asymmetric: its bandwidth shrank as the gain increased. The shift in wavelength of maximum spontaneous emissions (i.e. the peak gain wavelength) to a shorter wavelength with pumping was also observed, as with Crowe and Ahearn's results [32]. They predicted from their results that noise should be reduced as the amplifier is driven close to saturation. For amplifiers operating under resonant conditions, a signal gain approaching 1000 (30 dB) could be obtained with an injection current of 500 mA (cf. the range of injection currents in homojunction SLAs). It was found that the signal gain was extremely sensitive to fluctuations in parameters like reflectivities under resonant operations. Again, gain saturation was observed in their experiments. Finally, although satisfactory amplification could be achieved at room temperature with a fairly low injection current, their results showed that their amplifier was extremely sensitive to temperature, an issue that has received similar concern in recent years [38, 39].

The development of SLAs gathered pace after the invention of heterostructures. Many experiments were performed, e.g. by the NTT group in Japan led by Mukai and Yamamoto [40–42], by Simon in France [38] and by the BTRL group in England [43]. Theoretical modelling of SLAs became more sophisticated, e.g. analytical models proposed by Mukai and Yamamoto [41], Marcuse [44], Adams *et al.* [45], Buus and Plastow [46], and a recent one proposed by Lowery [47]. Experimental work on system applications with SLAs has also proceeded rapidly in recent years, as shown in Table 3.3. SLAs have reached a high degree of sophistication and a systematic classification and discussion of the various types of SLAs is now possible as compared with the

Table 3.3 Recent transmission experiments with (a) fibre amplifiers and (b) semiconductor laser amplifiers

Year	Lab.	Bit rate (Gbit/s)	Distance (km)	Comments
1989	NTT	1.8	212	Booster + pre-amp used
1989	BTRL	0.565	—	DPSK system
1989	KDD	1.2	267	2 amplifiers used
1989	NTT	20	—	Soliton transmission (ST)
1989	BellCore	11	260	2 amplifiers used
1989	Fujitsu	12	100	—
1989	KDD	1.2	904	12 amplifiers used
1990	NTT	2.5	2223	25 amplifiers used
1991	BTRL	2.5	10 000	Recirculating loop
1991	NTT	10	1 000 000	ST, 12 amplifiers used
1991	NTT	20	500	Soliton transmission
1992	NTT	2.4	309	4 repeaters + pre-amp.
1992	NTT	10	309	4 repeaters + pre-amp.

relative lack of diversity of types of SLAs in the early 1970s. In the next section we will assess the present state of the art for the different types of SLAs that are now available.

3.4   TYPES OF SEMICONDUCTOR LASER AMPLIFIERS

SLAs can be classified by their differences in operation, or by differences in their structures. The following discussion will cover both of these methods of classification.

3.4.1   Operational classification

Because any SLA is principally a structure derived from a semiconductor laser diode (SLD), we can generalise the structure of a SLA by an embedded active region with two end-facets for the input and output of light signals (Figure 3.9) [2]. The various structural details will be discussed in subsection 3.4.2. With this simplified model, we can classify SLAs into two broad categories.

As discussed in section 3.2, the end-facets with power reflection coefficients R_1 and R_2 (Figure 3.9) are responsible for providing optical feedback into the amplifier. These finite reflections arise because of the discontinuities of the refractive index between the semiconductor and air [49, 50]. In a semiconductor laser such reflections provide the necessary feedback F to excite oscillations when the single pass gain G_s satisfies the threshold condition of $FG_s = -1$, as depicted by equation (3.1). Because we are interested in linear amplification, the value of G_s can be reduced by reducing the injection current below the threshold (as with the early experiments performed by Crowe and Craig [31] and Crowe and Ahearn [32], such that $|FG_s| < 1$). In this case the amplifier becomes a *Fabry–Perot amplifier* (FPA). It can be seen that because of the finite

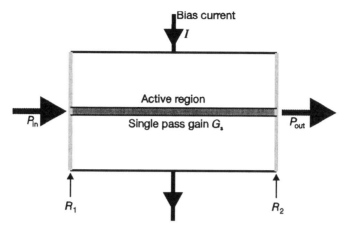

Figure 3.9 A simplified schematic of a semiconductor laser amplifier

reflectivities of the end-facets, some resonant cavity properties will be observed. As shown in Chapter 4, the resonating characteristics of the cavity will modulate the material gain spectrum introducing ripples along the amplifier gain spectrum [41] (see Figure 3.10). The level of ripples will affect the bandwidth of the SLA [13–51]. If the ripple level exceeds 3 dB, the bandwidth of the amplifier will be reduced to that of a single longitudinal cavity mode. As the theoretical analysis on FPAs shows (see Chapter 5), the gain ripple of the amplifier actually depends on the proportion of feedback F [6]. Hence the level of ripples can be minimised by reducing the values of the reflection coefficient of the end-facets.

A traditional way to reduce facet reflectivities is by depositing dielectric coatings, or anti-reflection (AR) coatings, onto the laser facets [52–55]. Ideally, if the coated laser facets have zero reflectivities (i.e. when the refractive index of

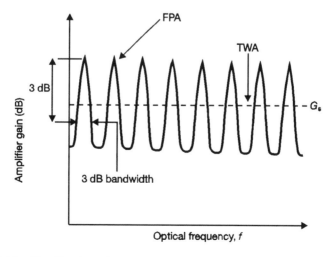

Figure 3.10 Classification of semiconductor laser amplifiers by their gain spectra

Table 3.4 Comparison between travelling-wave and Fabry–Perot amplifiers

Amplifier characteristics	TWA	FPA
Amplifier gain	20 dB ($R = 0.1\%$)	20–20 dB
	30 dB ($R = 0.01\%$)	
Gain bandwidth	20 THz	8–43 GHz
Saturation output level	5–10 dBm	-7 to -12 dBm
Noise	Low signal–spontaneous beating; high spon–spon beating	High signal–spontaneous beating; low spon–spon beating
Applications:		
Pre-amplifiers	◇ (with filter)	◇
Repeaters	◇ (with filter)	◇
Booster amplifiers	◇	
Post-amplifiers	◇	

the semiconductor material appears to be matched with air), the ripples will disappear, and the gain spectrum becomes more uniform with a wider bandwidth (Figure 3.10) [56]. An amplifier with no gain ripple is known as a *travelling-wave amplifier* (TWA).

A comparison between the gain, bandwidth, saturation output power and noise characteristics of these SLAs can be found in Table 3.4 [13, 38, 51]. A detailed treatment of the differences in performance between TWAs and FPAs will be found in subsequent chapters. However, it is useful to have some qualitative account at this stage for clarification.

FPAs have a very high gain when the wavelength of the input signals matches the peak transmission wavelength of (i.e. tuned to) the cavity resonance [37]. However, the amplifier gain is extremely sensitive to the structural and operational parameters of FPAs [38]. The signal gain in TWAs is more stable than that of FPAs, and is less sensitive to structural parameters and operational conditions [23]. Because of the presence of cavity resonance in FPAs, the ripples along the amplifier gain spectrum reduce their bandwidth. Hence TWAs have a wider bandwidth than FPAs, although the narrow bandwidth of FPAs is useful for active optical filters [2] and non-linear switching purposes [13]. Because of the presence of end-facet reflectivities, photons travel back and forth continuously across a FPA due to reflections. Some will be transmitted as output, but some will keep on travelling inside the cavity, generating more stimulated emissions from the population inversion. This enhances the pumping action and hence a FPA will saturate faster than a TWA without end reflections (where photons only travel across the cavity once). The output saturation power is therefore higher in TWAs. With respect to noise, the resonating action in FPAs can filter off some spontaneous emission power, especially that due to beating between spontaneous emissions [1]. However, the high gain due to cavity resonances in FPAs can enhance the beating between signal and spontaneous emission in the amplifier [42]. As we will see later, this can increase the overall noise figure in a FPA. Therefore the overall noise figure is lower in a TWA compared with a FPA [104].

In general, it can be seen that TWAs appear to be more promising than FPAs, perform closer to the ideal, and are found to be more useful in many applications [6]. Indeed, much research effort has been devoted recently to developing and fabricating TWAs [13, 23, 38, 51]. In practice, TWAs with zero end-facet reflectivities are impossible to fabricate. This is because traditional AR coatings cannot reduce the facet reflectivities to absolute zero [57–60]. Residual reflectivities still exist, and ripples are still present in the amplifier gain spectrum [38]. O'Mahony [13] used the term *near travelling-wave amplifiers* (NTWAs) to describe these real devices, which have a characteristic lying between that of a FPA and an ideal TWA. The criteria for NTWAs is that the ripple level in the amplifier gain spectrum should be less than 3 dB. Nearly all fabricated devices which are coated to reduce end-facet reflectivities are in fact NTWAs.

In addition to the residual reflectivities in AR coatings, there is a further difficulty in employing these coatings to eliminate facet reflectivities. No matter whether the AR coatings are single layered [59] or multi-layered [61, 62] they will have a minimum reflectivity for a finite range of optical wavelengths only. This so-called *window* of minimum reflectivity for the AR coatings is difficult to match with the peak gain region of the material gain spectrum [38, 63]. Even if the wavelength for minimum reflectivity of the AR coatings is *designed* to match the peak gain wavelength of the material, it is difficult to *achieve* the design during the actual fabrication process [51]. A further complication is that in many TWA applications a large input signal bandwidth is involved. The corresponding residual reflectivities of the end-facets due to the coatings are different for different wavelengths of the input signal. Hence, the result is a difference in signal gain for different wavelengths. Whilst some of the wavelengths of the input are amplified more, some will be amplified less, thereby distorting the output signals. Such behaviour is far from that of ideal TWAs, because the amplifier gain spectrum should be uniform along the pass band in a TWA. Recently, it was discovered that the roughness that remains on the facets of the amplifier due to coatings (which is always finite because of limitations in the fabrication process) will also affect the values of their facet reflectivities [64].

In order to overcome these problems, other techniques are proposed to reduce facet reflectivities [2]. Recent techniques include using a window region [65], a buried facet structure [66] and an angled facet structure (i.e. tilted end-facets) [67–69]. Some work on minimising end-facet reflectivities in semiconductor waveguides for other opto-electronic devices may also be used in SLA fabrication [70, 71].

3.4.2 Structural classification

In the previous subsection the structural details inside the SLA have been ignored by using the simplified model of Figure 3.9. In real devices there are several possible types of structural layout for SLAs. Some of them are shown in Figure 3.11.

The simplest structure is a buried heterostructure (BH) (Figure 3.11(a)), which is the most common structure found in SLDs as well as SLAs [37]. The active region is embedded in layers of semiconductors (i.e. in *stratified* media) to improve the confinement of light. Further improvement in the performance of SLAs can be obtained by using separate confinement heterostructures

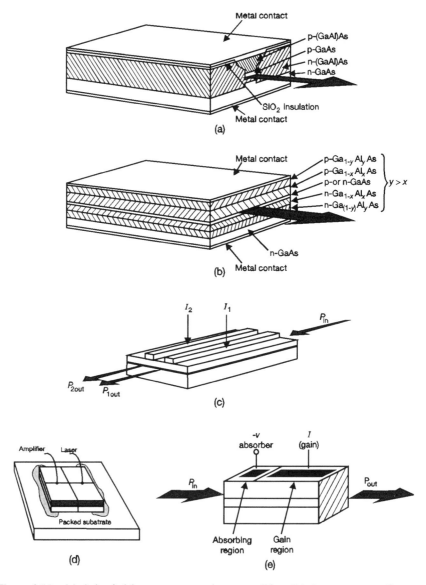

Figure 3.11 (a) A buried heterostructure laser amplifier. (b) A separate confinement heterostructure laser amplifier. (c) A twin-guided laser amplifier. (d) A cleaved substrate laser amplifier. (e) A split-contact laser amplifier

(SCH) (Figure 3.11(b)) [14], in which the carriers and light are separately confined (in contrast to BH, where both the light and carriers are confined within the active region [5]). Depending on the exact fabrication procedure, variations of this basic structure exist. An extensive review can be found in reference [4].

There are also SLA structures that are very different from this basic single waveguide, single active region structure. A twin-guided laser amplifier (TGLA) Figure 3.11(c) is a SLA with two waveguides, which was proposed for the integration of a laser with an amplifier/detector for injection locked amplifier (ILA) applications [6, 72]. This structure is found to be promising in active directional couplers and switches [73, 74]. A cleaved substrate laser amplifier (CSLA) (Figure 3.11(d)) usually consists of an oscillator and an amplifier separated by a small cleaved gap [34], which is again useful for ILA applications (e.g. de-modulation of FM signals [2]). Recently Marshall *et al.* [43] reported non-linear switching applications using a split contact laser amplifier (SCLA) (Figure 3.11(e)). It should be noted that for all of these novel structures the structure of a SLA has been refined for different applications, and the resulting structure is very different from an ordinary BH SLA.

3.5 RADIATIVE TRANSITION IN SEMICONDUCTORS

We have seen in Chapter 2 that in an optical amplifier at a finite temperature, random spontaneous emission events will occur. Einstein argued that to satisfy Planck's law of spectral distribution of electromagnetic radiation, stimulated emissions must also occur, which will manifest themselves as an optical gain if the amplifier is under negative temperature or population inversion [105]. This treatment has been considered in some detail in Chapter 2 for a simple two-level system. In semiconductors, a similar radiative transition of carriers occurs. However, instead of occurring between two discrete and distinct energy levels, as postulated in Chapter 2, in semiconductors these radiative transitions will occur between two energy bands known as conduction and valence bands, respectively. The basics of semiconductor band theory is covered in many textbooks [5, 7]. The reader is recommended to refer to them for more fundamental questions about the definition of Fermi levels, the meaning of density of states, etc. In this section we give a very brief outline of the relevant equations for radiative transitions in a semiconductor where bands have exponential band tailing [5, 22]. Such band tailings occur because of the presence of randomly distributed impurities in the material due to heavy doping [4].

3.5.1 Stimulated emissions

It can be shown that by considering the density of states and the population distribution of holes and electrons in a parabolic band, as shown in Figure 3.12, the net stimulated emissions will occur if and only if [3, 4]

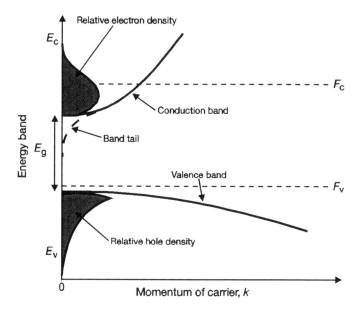

Figure 3.12   Energy band versus momentum of carrier

$$\exp\left(\frac{E_c - hf - F_v}{kT}\right) > \exp\left(\frac{E_c - F_c}{kT}\right) \tag{3.16}$$

where E_c is the energy of a state in the conduction band, F_c and F_v are the quasi-Fermi levels in the conduction and valence bands, respectively, h is Planck's constant, f is the optical frequency, k is the Boltzmann constant and T is the temperature of the semiconductor. The above equation can be simplified to

$$F_c - F_v > hf \tag{3.17}$$

which indicates the condition for a net positive value of stimulated emission, i.e. maintaining the population inversion. Under such conditions the optical gain of the material can be found from the stimulated emission rate, and the corresponding spontaneous emission can be found from the spontaneous emission rate. The stimulated emission rate can be calculated if the transition probability B_{21} for the stimulated transition is known. To calculate B_{21} we must consider the mechanisms involved when the electrons in the semi-conductor interact with the radiation field. This requires a knowledge of the quantum mechanics of the system, involving the wave functions of the electron. The appropriate approach is to use time-dependent perturbation of the Schrodinger equation, such that the effects of the incident radiation on the system can be expressed in terms of the wave functions that describe the system in the absence of radiation [106]. We will simply quote the result here [5] as

$$B_{21} = \frac{e^2 h |M|^2}{2m^2 \epsilon_0 \mu^2} \tag{3.18}$$

where e and m are the charge and mass of an electron and M is the matrix element of the momentum operator connecting the wave function of the upper state in the conduction band to that of the lower state in the valence band [106]. Evaluation of the matrix element M involves finding the appropriate wave functions that describe the conduction and valence band states of the semiconductor. This in turn requires a knowledge of the band structure of the material.

For transitions under k, selection of the polarisation of the wave functions during transition is taken into account. Under these circumstances it can be shown that the matrix element M requires the wave vector for wave functions during transitions to be identical [5]. In this situation, the value of M can be found approximately as

$$|M|^2 = \frac{m^2 E_g[1 + \Delta/E_g](1 - m_c/m)}{3m_c[1 + 2\Delta/3E_g]} \tag{3.19}$$

where m is the free electron mass, m_c is the effective mass of the electron in the conduction band, E_g is the band gap energy and Δ is the spin–orbital splitting. Using this expression, the transition probability is given by

$$B_{21} = \frac{e^2 h[1 + \Delta/E_g](1 - m_c/m)}{6m_c \epsilon_0 \mu^2[1 + 2\Delta/3E_g]} \tag{3.20}$$

With the value of B_{21} known, the material gain of the amplifying medium can be found following the procedure described in reference [28].

3.5.2 Spontaneous emissions

In general, the transition probability of spontaneous emission can be expressed in terms of B_{21} (see section 2.2 of Chapter 2). Then the spontaneous emission recombination rate can be found from the following expression [5]:

$$R_{sp} = \int_{-\infty}^{-(b+a)} \int_{-b}^{\infty} Z(E_c - E_v)B(E_c, E_v)V\rho_c(E_v)\rho_v(E_c)f_c(1 - f_c)dE_c\,dE_v \tag{3.21}$$

where $Z(E)$ indicates the average number of modes in the radiation field with photon energy E, V is the volume of the material, E_c and E_v are energy of states within the conduction and valence bands, respectively (see Figure 3.12), f_c and f_v are the occupational probabilities of the state in the conduction and valence bands at energies E_c and E_v, respectively, and ρ_c and ρ_v are the density of states at the conduction and valence bands, respectively. The constants a and b are chosen such that the integration terminates within the forbidden bandgap. Notice that $B = B_{21}$. The above equation can be formulated by considering the number of transitions across the bands per second, similar to our analysis in the two-level system in Chapter 2. We can rewrite the above equation in the following form:

$$R_{sp} = Z \int_0^n \int_0^p B(n_\rho, p_\rho) V \, dn \, dp \tag{3.22}$$

where the variables n_ρ and p_ρ are alternatives to E_c and E_v, which represent the total electron and hole states (whether they are occupied or not) up to energy E_c and down to energy E_v. Z is taken outside the integral as being approximately constant over the small range of photon energies involved in the integration. Usually, the above form of B can be handled in two situations [5]: first the non-inverted Boltzmann condition with both quasi-Fermi levels lying inside the bandgap; and secondly when B can be approximated to be independent of the carrier densities n and p. Hence the spontaneous emission rate is given approximately by

$$R_{sp} \approx Bnp \tag{3.23}$$

which is the well-known expression used throughout this book and other references in considering the radiative recombination rate [4, 27].

3.6 APPLICATIONS OF SEMICONDUCTOR LASER AMPLIFIERS

We will conclude this chapter by discussing various applications of SLAs. References [2] and [13] have given extensive reviews of various applications of SLAs. We saw in Chapter 2 that the exact performance requirements of SLAs vary with their application. They are summarised in Table 3.5 and will be discussed below. In addition, the problems with the present design of SLAs for different application requirements will also be introduced.

3.6.1 Non-regenerative repeaters

Conventionally, the in-line repeaters in optical fibre communications are *regenerative* (Figure 3.13(a)). They are regenerative because the amplified signals at the output of the repeaters are actually regenerated signals. This is illustrated in Figure 3.13(a). The whole process involves the detection of a weak incoming signal using a photodetector (PD), which converts the light

Table 3.5 Performance requirements in amplifier applications

Performance requirements	Pre-amplifier	In-line repeater	Booster amplifier
1. Sufficient small signal gain	Required	Required	Required
2. Wide gain bandwidth	Required	Required	Required
3. Polarisation-insensitive signal gain	Required	Required	Required
4. High saturation output power	—	Indispensable	Indispensable
5. Small noise figure	Indispensable	Indispensable	Indispensable
6. Use of narrow-band optical filter	Indispensable	Required	Required

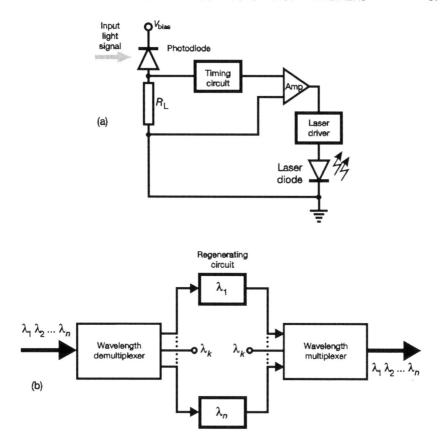

Figure 3.13 (a) Schematic of a regenerative repeater. (b) Regenerative repeaters for WDM systems

signal into an electrical current, and is then fed into an electronic amplifier for amplification. The amplified current is then used to drive a SLD and hence regenerates the required signal. Such repeaters are not only expensive but also suffer from poor sensitivity at the input to PDs, and noise is an inherent problem [2]. In addition, the laser driver electronics have to be designed for different schemes of modulation and signal shaping, and hence their versatility is very limited. It can also be expected that maintenance is difficult for these repeaters, especially when they are under the sea [75]. Reliability is also a problem.

Additional problems arise when regenerative repeaters are used in a wavelength division multiplexed (WDM) system (Figure 3.13(b)). In a WDM system the input to the repeaters consists of signals of several different wavelengths (or channels). Hence each channel requires a distinct PD–electronic amplifier–SLD path, resulting in arrays of PDs and SLDs as well as complex electronic circuits in the package. The reliability and ease of maintenance of this type of repeater is of course very poor, and adaptability will be a major problem when the WDM system has to be upgraded for different modulation schemes or speed.

SLAs can be used to replace regenerative repeaters. Because SLAs will directly amplify the weak incoming optical signals without any regeneration process, in this application SLAs are known as *non-regenerative* repeaters. They do not suffer from the problem of adaptability because there is no need for modulation-dependent electronics. For NTWAs with a wide bandwidth (> 4000 GHz), simultaneous amplification of several channels in a WDM system is possible [76, 77]. Because the amplification process is independent of the direction of transmission in a SLA, bi-directional transmission is also possible when SLAs are used as in-line repeaters [78], and hence WDM systems with bi-directional transmission become feasible.

When real devices are used in this application, high output saturation power and low noise figures are essential to avoid saturating the repeater chain by amplified spontaneous emissions (ASE) [1]. A wide bandwidth is desirable for WDM applications, but channel cross-talk in the amplifiers is a constraint that limits the maximum number of channels that can be transmitted through the system [79]. In bi-directional, multi-channel transmissions, non-linear effects like non-degenerate or nearly degenerate four-wave mixing can occur in the SLA, generating spurious frequencies at the amplifier output [77, 80, 81], which will degrade the performance of the system especially when coherent detection is used to detect the signals. The channel cross-talk and non-linear effects in SLAs have also to be investigated and minimised to improve the performance of SLAs as repeaters in WDM systems.

3.6.2 Pre-amplifiers to optical receivers

In optical communications, receivers are required to detect the optical signals transmitted along the fibre and to convert them back to electrical signals. A typical configuration of an optical receiver is shown in Figure 3.14, which consists of a PD and a FET amplifier [82]. Using an avalanche photodetector (APD), the largest gain bandwidth which can be obtained from the receiver is ~ 70 GHz [3]. For even wider bandwidths, e.g. for a subcarrier multiplexed lightwave system for multi-channel video distribution [83], an optical pre-amplifier will be needed [35]. When a SLA is used as a front-end pre-amplifier to the receiver, it can amplify the incoming weak signal, thereby improving the sensitivity of the receiver [84, 85]. Usually, when a SLA is incorporated into the receiver, a high-speed *pin* photodetector will be used to convert the light signals into electrical signals, and a receiver bandwidth as wide as the material gain bandwidth of the SLA can be achieved [86, 87]. For this particular application, SLAs with a wide bandwidth and low noise figure are required [15]. TWAs are ideal candidates in this case, although optical filtering may be needed to eliminate the beating noise between spontaneous emissions [6]. In addition, a high coupling efficiency between the SLA and the *pin* PD, as well as between the SLA and the optical fibre, is required as the improvement in receiver sensitivity depends strongly on the coupling loss [2]. Therefore reliable TWAs,

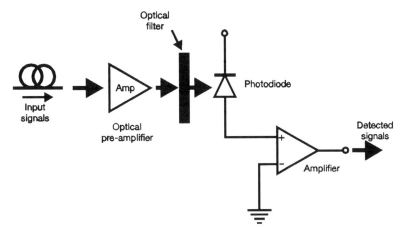

Figure 3.14 Configuration of an optical receiver

as well as a high coupling efficiency between the optical components, are the major targets to be achieved if SLAs are to be used effectively as pre-amplifiers to optical receivers.

3.6.3 Bistable and switching applications

The phenomenon of optical bistability in semiconductor lasers has received much attention in the development of semiconductor lasers [45]. There are two distinctive ways to operate SLAs as a bistable element [2, 3], i.e. electrically or optically. Because a SLA absorbs all the input light power when there is no injection or bias current (i.e. no pumping), and it provides gain to the input signals when the current is fully turned on, it can be used as a lossless gating switch [7]. In this application the SLA will block all incoming signals when there is no electrical current flowing to it (i.e. the "off" state), whereas when the current flowing across it is fully turned on (i.e. the "on" state), the SLA can pass the signals with gain to compensate losses in the optical fibre network. Such gating switches are ideal for optical routing in an optical fibre network. A SLA can also be switched optically as well, due to the asymmetric gain spectrum of semiconductors. It has been shown by many experiments that as the input power increases, the wavelength of maximum output power in a FPA shifts to the red side of the optical spectrum [13]. Such a detuning phenomenon can be used to provide optical bistability. This can be achieved by enclosing the amplifying medium in a highly resonating Fabry–Perot cavity [45, 89, 90]. Usually, we can bias a FPA to just below its oscillation threshold to achieve this. The input signal wavelength is slightly offset from that of the cavity resonance to a larger value. As the input power increases, the cavity resonance shifts to a longer wavelength, and the output power increases. This is illustrated in Figure 3.15(a). The transfer characteristics of P_{out} versus P_{in} of this bistable amplifier are shown in Figure 3.15(b). In practice, switching occurs

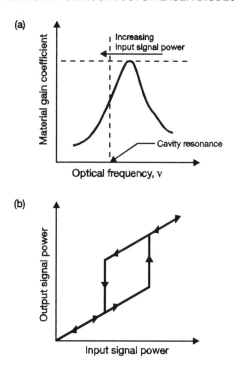

Figure 3.15   (a) The detuning of a FPA with input signal power. (b) Resulting transfer characteristics of the bistable amplifier

at relatively low input power because the SLA is highly resonating and is biased just below the oscillation threshold [13]. Moreover, as the input optical power is increased and then reduced, a hysteresis loop in the output power is observed, as shown in Figure 3.15(b). This hysteresis can be enhanced by introducing an additional absorption region at the input, resulting in an optical bistable flip-flop element [2]. By using a twin-guide structure, it is possible to actually construct optical logic gates with two input channels and to utilise this optical bistability. Indeed, this is the principle of an active optical directional coupler [74, 91]. Additional applications of optical bistability can be found in pulse shaping [13]. The generation and shaping of optical pulses with a very short duration can be achieved by utilising the properties of optical bistability, which are extremely useful in high-speed pico- and femtosecond optics [14].

In contrast to the previous two applications, when SLAs have to be used as switches, they must be highly resonating with a narrow bandwidth [13]. Hence FPAs become ideal candidates in this application. Although promising, at present the applications of SLAs as switches and bistable amplifiers are hindered by relatively low operational speed and poor stability. These drawbacks arise because of the finite gain recovery time due to carrier recombination mechanisms. In order to improve the bistable performance of

SLAs, the gain recovery mechanisms in SLAs should be understood more fully, and some structural refinements may be necessary [2].

3.6.4 Other applications

In addition to the above applications, SLAs have been found to be potential devices in many other applications. For example, SLAs can be used as modulators [92]. The modulation of optical signals can be achieved by modulating the refractive index in the active medium [93]. In SLAs this can be achieved very easily by changing the injection current. Since the material gain changes with current, the refractive index in the SLA changes as predicted by the Kramers–Kronig relation [7]. If we modulate the injection current of the SLA sinusoidally, frequency modulation will result in the optical signals. This makes the SLA a very convenient form of external optical modulator, which can be integrated easily with semiconductor lasers to form a high-speed transmitter [94].

Another application of the SLA is to use it as a detector/monitor in optical fibre networks [72, 95]. As optical signals pass through the amplifier, stimulated emissions, which occur due to the material gain in the active region, will draw carriers from the population inversion (see equations (2.4) and (2.7) in Chapter 2). If the optical signals are intensity modulated, the fluctuations in the optical intensity due to modulation will induce fluctuations in the injection current as well. These fluctuations in the current can then be amplified and detected. In this way the SLA becomes a detector. Although it may not be as sensitive as an ordinary photodiode, the possibility of detecting signals using SLAs makes them suitable as monitors along the transmission span of an optical fibre communication system [96, 97]. These monitors will detect the intermediate optical power within the transmission span, and if this power gets too low, it indicates that there is probably a break in the preceding transmission path and that appropriate maintenance steps should be undertaken. In addition, such monitors can be used to detect control protocols which are sent along with the optical signals in an optical network, which will be used to control the routing of optical signals.

Other recent proposed applications of SLAs include: use as a component in high-flexibility optical filters [98], as an injection-locked amplifier for the demodulation of FM signals [12], as a component in active optical switching matrices [99], as a channel dropping node [100], in optical phase-locked-loops [101] and as a dual-function gate/amplifier in an optical fibre network [102]. The reader might ask why, in spite of the endless list of potential applications, the development of SLAs seemed to be much slower in comparison with that for SLDs. Eisenstein [2] considered this to be the result of the slow speed of development in fabrication technology as well as the theoretical design of AR coatings. Indeed, even after the problem of reducing facet reflectivities has finally been overcome, there are many other problems remaining. The reasons behind the occurrence of these problems and the possible solutions will be examined in subsequent chapters.

3.7   REFERENCES

[1] Y. Yamamoto and T. Mukai, "Fundamental of optical amplifiers", *Opt. Quantum Electron.*, **QE-21**, pp. S1–S14, 1989.

[2] G. Eisenstein, "Semiconductor optical amplifiers", *IEEE Circuits and Devices Mag.*, pp. 25–30, 1989.

[3] A. Yariv, *Optical Electronics*, 3rd edition, Holt–Saunders, 1985.

[4] G. P. Agrawal and N. K. Dutta, *Long-wavelength Semiconductor Lasers*, Van-Nostrand Reinhold, New York, 1986.

[5] G. H. B. Thompson, *Physics of Semiconductor Laser Devices*, John Wiley and Sons, New York, 1980.

[6] Y. Mukai, Y. Yamamoto and T. Kimura, "Optical amplification by semiconductor lasers", *Semiconductors and Semimetals*, **22**, Part E, pp. 265–319, Academic Press, London, 1985.

[7] A. Yariv, *Quantum Electronics*, 3rd edition, John Wiley and Sons, New York, 1989.

[8] M. J. Adams, *An Introduction to Optical Waveguides*, John Wiley and Sons, New York, 1981.

[9] D. Marcuse, *Theory of Dielectric Optical Waveguides*, Academic Press, London, 1974.

[10] A. W. Snyder and J. D. Love, *Optical Waveguide Theory*, Chapman & Hall, London, 1983.

[11] M. Cross and M. J. Adam, "Waveguiding properties of stripe-geometry double heterostructure injection lasers", *Solid State Electron.*, **15**, pp. 919–921, 1972.

[12] C. Y. J. Chu and H. Ghafouri-Shiraz, "Structural effects on polarization sensitivity of travelling-wave semiconductor laser amplifiers", *3rd Bangor Communication Symposium*, United Kingdom, pp. 19–22, 1991.

[13] M. J. O'Mahony, "Semiconductor laser optical amplifiers for use in future fiber systems", *IEEE J. Lightwave Technol.*, **LT-6**, No. 4, pp. 531–544, 1988.

[14] T. Saitoh and T. Mukai, "Structural design for polarization-insensitive travelling-wave semiconductor laser amplifiers", *Opt. Quantum Electron.*, **21**, pp. S47–S48, 1989.

[15] A. G. Failla, G. P. Brava and I. Montrosset, "Structural design criteria for polarization insensitive semiconductor optical amplifiers", *J. Lightwave Technol.*, **LT-8**, No. 3, pp. 302–308, 1990.

[16] H. Ghafouri-Shiraz and C. Y. J. Chu, "Refractive index control on the polarization sensitivity of semiconductor travelling-wave laser amplifiers", *Microwave and Opt. Technol. Lett.*, **5**, No. 3, pp. 152–154, 1992.

[17] H. Ghafouri-Shiraz and C. Y. J. Chu, "Analysis of waveguiding properties of travelling-wave semiconductor laser amplifiers using perturbation technique", *Fiber and Integrated Optics*, **11**, pp. 51–70, 1992.

[18] H. Ghafouri-Shiraz and C. Y. J. Chu, "Analysis of polarization sensitivity in semiconductor laser amplifiers using the perturbation method", *Trans. IEICE Jpn, National Convention Record*, Paper No. C-188, 1991.

[19] G. Lasher and F. Stern, "Spontaneous and stimulated recombination radiation in semiconductors", *Phys. Rev.*, **133**, No. 2A, pp. 553–563, 1964.

[20] C. J. Hwang, "Properties of spontaneous and stimulated emissions in GaAs junction laser; I & II", *Phys. Rev. B*, **2**, No. 10, pp. 4117–4134, 1970.

[21] F. Stern, "Calculated spectral dependence of gain in excited GaAs", *J. Appl. Phys.*, **47**, No. 2, pp. 5382–5386, 1976.

[22] M. J. Adam, "Theoretical effects of exponential band tails on the properties of the injection laser", *Solid State Electron.*, **12**, pp. 661–669, 1969.

[23] T. Saitoh and M. Mukai, "1.5 μm GaInAs travelling-wave semiconductor laser amplifier", *IEEE J. Quantum Electron.*, **QE-23**, No. 6, pp. 1010–1020, 1987.

[24] H. Ghafouri-Shiraz, "A model for peak-gain coefficient in InGaAsP/InP semiconductor laser diodes", *Opt. Quantum Electron.*, **20**, pp. 153–163, 1988.

[25] H. Ghafouri-Shiraz, "Analysis of characteristics of long-wavelength semi-conductor laser diode amplifiers", *Opt. Quantum Electron.*, **19**, pp. 303–311, 1987.

[26] J. Wang, H. Olesen and K. E. Stubkjaer, "Recombination, gain and bandwidth characteristics of 1.3 μm semiconductor laser amplifiers", *IEEE J. Lightwave Technol.*, **LT-5**, No. 1, pp. 184–189, 1987.

[27] R. Olshansky, C. B. Su, J. Manning and W. Powazink, "Measurement of radiative and nonradiative recombination rates in InGaAsP and AlGaAs light sources", *IEEE J. Quantum Electron.*, **QE-20**, No. 8, pp. 838–854, 1984.

[28] M. Osinski and M. J. Adam, "Gain spectra of quaternary semiconductors", *IEE Proc.*, **129**, Part J, No. 6, pp. 229–236, 1982.

[29] I. D. Henning, M. J. Adams and J. V. Collins, "Performance prediction from a new optical amplifier model", *IEEE J. Quantum Electron.*, **QE-21**, No. 6, pp. 609–613, 1985.

[30] T. Mukai, K. Inoue and T. Saitoh, "Homogeneous gain saturation in 1.5 μm InGaAsP travelling-wave semiconductor laser amplifiers", *Appl. Phys. Lett.*, **51**, No. 6, pp. 381–383, 1987.

[31] J. W. Crowe and R. M. Craig, "Small-signal amplification in GaAs lasers", *Appl. Phys. Lett.*, **4**, No. 3, pp. 57–58, 1964.

[32] W. J. Crowe and W. E. Ahearn, "Semiconductor laser amplifier", *IEEE J. Quantum Electron.*, **QE-2**, No. 8, pp. 283–289, 1966.

[33] W. F. Kosonocky and R. H. Cornely, "GaAs Laser amplifier", *IEEE J. Quantum Electron.*, **QE-4**, No. 4, pp. 125–131, 1968.

[34] M. B. Chang and E. Garmire, "Amplification in cleaved-substrate lasers", *IEEE J. Quantum Electron.*, **QE-16**, No. 9, pp. 997–1001, 1980.

[35] A. J. Arnaud, "Enhancement of optical receiver sensitivity by amplification of the carrier", *IEEE J. Quantum Electron.*, **QE-4**, No. 11, pp. 893–899, 1968.

[36] S. D. Personick, "Applications for quantum amplifiers in simple digital optical communication systems", *Bell Syst. Technol. J.*, **52**, No. 1, pp. 117–133, 1973.

[37] D. Schicketanz and G. Zeidler, "GaAs-double-heterostructure lasers as optical amplifiers", *IEEE J. Quantum Electron.*, **QE-11**, No. 2, pp. 65–69, 1975.

[38] J. C. Simon, "GaInAsP semiconductor laser amplifiers for single-mode fiber communications", *J. Lightwave Technol.*, **LT-5**, No. 9, pp. 1286–1295, 1987.

[39] M. S. Lin, A. B. Piccirilli, Y. Twu and N. K. Dutta, "Temperature dependence of polarization characteristics in buried facet semiconductor laser amplifiers", *IEEE J. Quantum Electron.*, **QE-26**, No. 10, pp. 1772–1778, 1990.

[40] T. Mukai and T. Yamamoto, "Noise characteristics of semiconductor laser amplifiers", *Electron. Lett.*, **17**, No. 1, pp. 31–33, 1981.

[41] T. Mukai and Y. Yamamoto, "Gain, frequency bandwidth, and saturation output power of AlGaAs DH laser amplifiers", *IEEE J. Quantum Electron.*, **QE-17**, No. 6, pp. 1028–1034, 1981.

[42] T. Mukai and Y. Yamamoto, "Noise in an AlGaAs semiconductor laser amplifier", *IEEE J. Quantum Electron.*, **QE-18**, No. 4, pp. 564–575, 1982.

[43] I. W. Marshall, O'Mahony, D. M. Cooper, P. J. Fiddyment, J. C. Regrault and W. J. Delvin, "Gain characteristics of a 1.5 μm non-linear split contact laser amplifier", *Appl. Phys. Lett.*, **53**, No. 17, pp. 1577–1579, 1988.

[44] D. Marcuse, "Computer model of an injection laser amplifier", *IEEE J. Quantum Electron.*, **QE-19**, No. 1, pp. 63–73, 1983.

[45] M. J. Adams, J. V. Collins and I. D. Henning, "Analysis of semiconductor laser optical amplifiers", *IEE Proc.*, **132**, Part J, pp. 58–63, 1985.

[46] J. Buus and R. Plastow, "A theoretical and experimental investigation of Fabry–Perot semiconductor laser amplifiers", *IEEE J. Quantum Electron.*, **QE-21**, No. 6, pp. 614–618, 1985.

[47] A. J. Lowery, "New in-line wideband dynamic semiconductor laser amplifier model", *IEE Proc.*, **135**, Part J, No. 3, pp. 242–250, 1988.

[48] T. Mukai, Y. Yamamoto and T. Kimura, "S/N and error rate performance in AlGaAs semiconductor laser preamplifier and linear repeater systems", *IEEE Trans. Microwave Theory Technol.*, **MTT-30**, No. 10, pp. 1548–1556, 1982.

[49] T. Ikegami, "Reflectivity of a mode at facet and oscillation mode in double-heterostructure injection laser", *IEEE J. Quantum Electron.*, **QE-8**, No. 6, pp. 470- 476, 1972.

[50] H. Ghafouri-Shiraz, "Facet reflectivity of InGaAsP BH laser diodes emitting at 1.3 μm and 1.55 μm", *Semiconduct. Sci. Technol.*, **5**, No. 2, pp. 139–142, 1990.

[51] T. Saitoh and T. Mukai, "Recent progress in semiconductor laser amplifiers", *J. Lightwave Technol.*, **LT-6**, No. 11, pp. 1156–1164, 1988.

[52] A. Hardy, "Formulation of two-dimensional reflectivity calculations based on the effective index method", *J. Opt. Soc. Am. A.*, **1**, No. 5, pp. 550–555, 1984.

[53] D. R. Kaplan and P. P. Deimel, "Exact calculation of the reflection coefficient for coated optical waveguide devices", *AT&T Bell Lab. Technol. J.*, **63**, No. 6, pp. 857–877, 1984.

[54] P. Kaczmarski, R. Baets, G. Franssens and P. E. Lagasse, "Extension of bi-directional beam propagation method to TM polarization and application to laser facet reflectivity", *Electron. Lett.*, **25**, No. 11, pp. 716–717, 1989.

[55] M. Yamada, Y. Ohmori, K. Takada and M. Kobayashi, "Evaluation of anti-reflection coatings for optical waveguides", *Appl. Opt.*, **30**, No. 6, pp. 682–688, 1991.

[56] C. A. Balanis, *Advanced Engineering Electromagnetics*, John Wiley and Sons, New York, 1989.

[57] R. H. Clark, "Theoretical performance of an anti-reflection coating for a diode laser amplifier", *Int. J. Electron.*, **53**, No. 5, pp. 495–499, 1982.

[58] R. H. Clarke, "Theory of reflections from antireflection coatings", *Bell Syst. Technol. J.*, **62**, No. 10, pp. 2885–2891, 1983.

[59] G. Eisenstein, "Theoretical design of single-layer antireflection coatings on laser facets", *AT&T Bell Lab. Technol. J.*, **63**, No. 2, pp. 357–364, 1984.

[60] T. Saitoh, T. Mukai and O. Mikami, "Theoretical analysis and fabrication of antireflection coatings on laser-diode facets", *J. Lightwave Technol.*, **LT-3**, No. 2, pp. 288–293, 1985.

[61] C. Vassallo, "Theory and practical calculation of antireflection coatings on semiconductor laser diode optical amplifiers", *IEE Proc.*, **137**, Part J., No. 4, pp. 193–202, 1990.

[62] C. Vassallo, "Some numerical results on polarization insensitive 2-layer antireflection coatings for semiconductor optical amplifiers", *IEE Proc.*, **137**, Part J., No. 4, pp. 203–204, 1990.

[63] J. Stone and L. W. Stulz, "Reflectance, transmission and loss spectra of multilayer Si/SiO₂ thin film mirrors and antireflection coatings for 1.5 μm", *Appl. Opt.*, **29**, No. 4, pp. 583–588, 1990.

[64] C. F. Lin, "The influence of facet roughness on the reflectivities of etched-angled facets for superluminescent diodes and optical amplifiers", *IEEE Photon. Technol. Lett.*, **4**, No. 2, pp. 127–129, 1992.

[65] I. Cha, M. Kitamura and I. Mito, "1.5 μm band travelling-wave semiconductor optical amplifiers with window facet structure", *Electron. Lett.*, **25**, No. 3, pp. 242–243, 1989.

[66] M. S. Lin, A. B. Piccirilli, Y. Twu and N. K. Dutta, "Fabrication and gain measurement for buried facet optical amplifier", *Electron. Lett.*, **25**, No. 20, pp. 1378–1380, 1989.

[67] C. E. Zah, C. Caneau, F. K. Shokoohi, S. G. Menocal, F. Favire, L. A. Reith and T. P. Lee, "1.3 μm GaInAsP near-travelling-wave laser amplifiers made by combination of angled facets and antireflection coatings", *Electron. Lett.*, **24**, No. 20, pp. 1275–1276, 1988.

[68] G. A. Alphonse, J. C. Connolly, N. A. Dinkel, S. L. Palfrey and D. B. Gilbert, "Low spectral modulation high-power output from a new AlGaAs

superluminescent diode/optical amplifier structure", *Appl. Phys. Lett.*, **55**, No. 22, pp. 2289–2291, 1989.

[69] D. Marcuse, "Reflection loss of laser mode from tilted end mirror", *IEEE J. Lightwave Technol.*, **LT-7**, No. 2, pp. 336–339, 1989.

[70] T. Baba and Y. Kokubun, "New polarization-insensitive antiresonant reflecting optical waveguide (ARROW-B)", *IEEE Photon. Technol. Lett.*, **1**, No. 8, pp. 232–234, 1989.

[71] K. W. Jelly and R. W. H Engelmann, "An etch tunable antireflection coating for the controlled elimination of Fabry–Perot oscillations in the optical spectra of transverse modulator structures", *IEEE Photon. Technol. Lett.*, **4**, No. 6, pp. 550–553, 1989.

[72] K. Kishino, Y. Suematsu, K. Utaka and H. Kawanishi, "Monolithic integration of laser and amplifier/detector by twin-guide structure", *Jpn J. Appl. Phys.*, **17**, No. 3, pp. 589–590, 1978.

[73] M. J. Adams, "Theory of twin-guides Fabry–Perot laser amplifiers", *IEE Proc.*, **136**, Part J, No. 5, pp. 287–292, 1989.

[74] M. J. Adams, D. A. H. Mace, J. Singh and M. A. Fisher, "Optical switching in the twin-guide Fabry–Perot laser amplifier", *IEEE J. Quantum Electron.*, **QE-26**, No. 10, pp. 1764–1771, 1990.

[75] T. Kimura, "Coherent optical fibre transmission", *IEEE J. Lightwave Technol.*, **LT-5**, No. 4, pp. 414–428, 1987.

[76] B. Glance, G. Eisenstein, P. J. Fitzgerald, K. J. Pollack and G. Raybon, "Optical amplification in a multichannel FSK coherent system", *Electron. Lett.*, **24**, No. 18, pp. 1157–1159, 1988.

[77] E. Dietrich, B. Enning, G. Grosskopf, L. Kuller, R. Ludwig, R. Molt, E. Patzak and H. G. Weber, "Semiconductor laser optical amplifiers for multichannel coherent optical transmission", *J. Lightwave Technol.*, **LT-7**, No. 12, pp. 1941–1955, 1989.

[78] D. J. Maylon and W. A. Stallard, "565 Mbit/s FSK direct detection system operating with four cascaded photonic amplifiers", *Electron. Lett.*, **25**, No. 8, pp. 495–497, 1989.

[79] G. Grosskopf, R. Ludwig and H. G. Weber, "Cross-talk in optical amplifiers for two-channel transmission", *Electron. Lett.*, **22**, No. 17, pp. 900–902, 1986.

[80] K. Inoue, T. Mukai and T. Saito, "Nearly degenerate four-wave mixing in a travelling-wave semiconductor laser amplifier", *Appl. Phys. Lett.*, **51**, No. 14, pp. 1051–1053, 1987.

[81] S. Ryu, K. Mochizuki and H. Wakabayashi, "Influence of non-degenerate four-wave mixing on coherent transmission system using in-line semiconductor laser amplifiers", *J. Lightwave Technol.*, **LT-7**, No. 10, pp. 1525–1529, 1989.

[82] H. Kressel, *Semiconductor Devices for Optical Communications*, 2nd edition, Springer-Verlag, Berlin, 1982.

[83] W. I. Way, C. Zah and T. P. Lee, "Application of travelling-wave laser amplifiers in subcarrier multiplexed lightwave systems", *IEEE Trans. Microwave Theory Technol.*, **38**, No. 5, pp. 534–545, 1990.

[84] K. Kannan, A. Bartos and P. S. Atherton, "High-sensitivity receiver optical preamplifier", *IEEE Photon. Technol. Lett.*, **4**, No. 3, pp. 272–274, 1992.

[85] K. Kikuchi, "Proposal and performance analysis of novel optical homodyne receiver having an optical preamplifier for achieving the receiver sensitivity beyond the shot-noise limit", *IEEE Photon. Technol. Lett.*, **4**, No. 2, pp. 195–197, 1992.

[86] Y. Yamamoto and H. Tsuchiya, "Optical receiver sensitivity improvement by a semiconductor laser amplifier", *Electron. Lett.*, **16**, No. 6, pp. 233–235, 1980.

[87] N. Nakagawa and S. Shimada, "Optical amplifiers in future optical communication systems", *IEEE LCS Mag.*, **1**, No. 4, pp. 57–62, 1990.

[88] T. Nakai, R. Ito and N. Ogasawara, "Asymmetric frequency response of semiconductor laser amplifiers", *Jpn J. Appl. Phys.*, **21**, No. 11, pp. L680-L682, 1982.

[89] A. J. Lowery, "Modelling spectral effects of dynamic saturation in semiconductor laser amplifiers using the transmission-line laser model", *IEE Proc.*, **136**, Part J, No. 6, pp. 320–324, 1989.

[90] C. T. Hultgren and E. P. Ippen, "Ultrafast refractive index dynamics in AlGaAs diode laser amplifiers", *Appl. Phys. Lett.*, **59**, No. 6, pp. 635–637, 1991.

[91] J. Singh, I. D. Henning, P. S. Mudhar, M. A. Fisher, S. Perrin, D. A. H. Mace and M. J. Adams, "A novel twin-ridge-waveguide optical amplifier switch", *IEEE Photon. Technol. Lett.*, **4**, No. 2, pp. 173–176, 1992.

[92] D. Bakewell, "Amplitude and phase modulation of laser amplifier using travelling wave model", *Electron. Lett.*, **27**, No. 4, pp. 329–330, 1991.

[93] A. Yariv and P. Yeh, *Optical Waves in Crystals*, John Wiley and Sons, New York, 1984.

[94] M. A. Ali, A. F. Elrefaie and S. A. Ahmed, "Simulation of 12.5 Gb/s lightwave optical time-division multiplexing using semiconductor optical amplifiers as external modulators", *IEEE Photon. Technol. Lett.*, **4**, No. 3, pp. 280–282, 1992.

[95] M. Gustvasson, L. Thylen and D. Djupsjobacka, "System performance of semiconductor laser amplifier detectors", *Electron. Lett.*, **25**, No. 2, pp. 1375–1377, 1989.

[96] R. M. Fortenberry, A. J. Lowery and R. S. Tucker, "Up to 16 dB improvement in detected voltage using two section semiconductor optical amplifier detector", *Electron. Lett.*, **28**, No. 5, pp. 474–476, 1992.

[97] K. T. Koai and R. Olshansky, "Simultaneous optical amplification, detection, and transmission using in-line semiconductor laser amplifiers", *IEEE Photon. Technol. Lett.*, **4**, No. 5, pp. 441–443, 1992.

[98] B. Moslehi, "Fibre-optic filters employing optical amplifiers to provide design flexibility", *Electron. Lett.*, **28**, No. 3, pp. 226–228, 1992.

[99] C. Burke, M. Fujiwara, M. Yamaguchi, H. Nishimoto and H. Honmou, "128 line photonic switching system using LiNbO₃ switch matrices and semiconductor travelling wave amplifiers", *J. Lightwave Technol.*, **10**, No. 5, pp. 610–615, 1992.

[100] C. Jørgensen, N. Storkfelt, T. Durhuus, B. Mikkelsen, K. E. Stubkjaer, B. Fernier, G. Gelly and P. Doussiere, "Two-section semiconductor optical amplifier used as an efficient channel dropping node", *IEEE Photon. Technol. Lett.*, **4**, No. 4, pp. 348–350, 1992.

[101] S. Kawamishi and M. Saruwatari, "10GHz timing extraction from randomly modulated optical pulses using phase-locked loop with travelling-wave laser-diode optical amplifier using optical gain modulation", *Electron. Lett.*, **28**, No. 5, pp. 510–511, 1992.

[102] K. T. Koai and R. Olshansky, "Dual-function semiconductor laser amplifier in a broad-band subcarrier multiplexed system", *IEEE Photon. Technol. Lett.*, **2**, No. 12, pp. 926–928, 1990.

[103] H. Ghafouri-Shiraz, "Temperature, bandgap-wavelength, and doping dependence of peak-gain coefficient parabolic model parameters for InGaAsP/InP semiconductor laser diodes", *IEEE J. Lightwave Technol.*, **LT-8**, No. 4, pp. 500–506, 1988.

[104] J. C. Simon, J. L. Favennec and J. Charil, "Comparison of noise characteristics of Fabry–Perot-type and travelling-wave-type semiconductor laser amplifiers", *Electron. Lett.*, **19**, No. 8, pp. 288–290, 1983.

[105] R. Loudon, *The Quantum Theory of Light*, 2nd edition, Oxford University Press, Oxford, UK, 1983.

[106] W. H. Louisell, *Quantum Statistical Properties of Radiation*, John Wiley and Sons, New York, 1973.

ANALYSIS OF TRANSVERSE MODAL FIELDS IN SEMICONDUCTOR LASER AMPLIFIERS (SLAs)

4.1   INTRODUCTION

In Chapter 3 we examined how optical amplification can be achieved in semiconductor lasers. We have seen that the formation of an active region due to population inversion in a forward biased p–n junction will introduce optical gain, and a simple dielectric waveguide will be formed due to the resulting refractive index step. The semiconductor laser amplifier amplifies signals that travel across the active region via the dielectric waveguide. In this chapter we apply some of the mathematical principles of electromagnetics to analyse how optical signals propagate across a semiconductor laser amplifier by treating them as electromagnetic waves [1, 2]. This investigation is important because the propagation characteristics of optical signals in the amplifier determine the electromagnetic field distribution inside the structure, which affects the photon density in the amplifier and is crucial in determining the amplifier gain, gain saturation and noise characteristics.

The electromagnetic analysis of a semiconductor laser amplifier (SLA) involves two steps. First, we have to determine the *transverse field distributions in the amplifier*. These are affected by the dimensions of the active region on the *x–y* plane of the dielectric optical waveguide formed in the active region [3, 4]. Secondly, we have to determine the *longitudinal field distributions*. These are affected by the amplifier length and the boundary conditions along the direction of propagation of the signals. Both of these field distributions in the amplifier depend on the group velocity of the signals, which is principally determined by the transverse field distributions. Furthermore, as we shall see later, several discrete and distinct field patterns can exist in a dielectric optical waveguide for a particular signal frequency. We designate each of these patterns as different *modes*, and each field pattern is known as a *modal field distribution*. The number of modes that can propagate across the SLA is determined by their group velocities and the dimensions of the waveguide (hence the transverse modal field distribution). Thus both the number and field patterns of the propagating modes of the signal in a particular SLA are determined ultimately by the transverse field distribution over its transverse x–y plane. In this chapter we concentrate on the analysis of the transverse modal fields in SLAs, which can help us to determine the propagation characteristics, or waveguiding characteristics, of the amplifier. These characteristics are extremely important as they will also affect the signal gain of the amplifier, which we will also examine in some depth in this chapter.

This chapter is organised as follows. Various methods of solving the transverse modal field distribution in a dielectric rectangular waveguide will be reviewed. We concentrate on rectangular structures as in most SLAs their active regions have a rectangular cross-section [5]. Solving this modal field distribution can determine two important parameters: the *effective index* [4, 6], and the *modal gain coefficient*, or the *effective gain*, of the amplifier [7]. The latter parameter can be used to study the polarisation sensitivity of the signal gain in a SLA [8, 9]. A method based on *perturbation analysis*, proposed by Ghafouri-Shiraz and Chu [10], will be used to calculate the modal gain coefficient of a SLA. This chapter is concluded by a discussion of the structural design implications for polarisation insensitive travelling-wave SLAs.

4.2 SOLUTION OF TRANSVERSE MODAL FIELDS IN RECTANGULAR OPTICAL WAVEGUIDES

The solution of transverse modal fields in a rectangular dielectric optical waveguide involves solving the wave equation for the structure. This topic has been covered in many textbooks on electro-optics, integrated optics or electromagnetics [1–4, 11, 12]. However, improvements in the technique for solving the wave equation in dielectric waveguides are still under active research. This is because to date no exact analytical solutions are available for this structure [13]. We first examine briefly the simplest case of a three-layer dielectric slab and the corresponding modal field solutions. We then review possible methods of solving the problem for more complex rectangular dielectric waveguides.

4.2.1 Solution for a three-layer slab (planar optical waveguide)

A schematic of the dielectric slab is shown in Figure 4.1. It is well known in electromagnetics that the electric and magnetic field of a source-free waveguide satisfies the homogeneous wave equation [2]. We can start to derive this equation from Maxwell's **curl** equations, which are

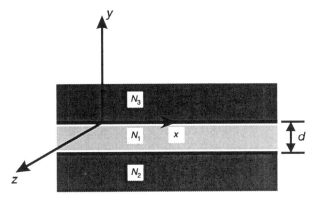

Figure 4.1   A planar three-layer dielectric slab

$$\nabla x E = -j\omega\mu H \tag{4.1}$$

$$\nabla x H = j\omega\epsilon E \tag{4.2}$$

The bold symbols represent vector quantities. The above equations are expressed in time harmonic form without current sources. In these equations, ϵ and μ are the relative permittivity and relative permeability of the medium, respectively. They can be spatially dependent (in an inhomogeneous medium), or even depend on the magnitude and direction of the electric and magnetic field vectors themselves (in an anisotropic medium). In the latter case, tensor analysis has to be used [14–16], which is beyond the scope of this book. In the following discussion we consider a homogeneous medium only.

By using vector identities, taking the **curl** of both sides of equation (4.1) and substituting equation (4.2) into it, we can arrive at the *scalar wave equation* for the scalar component of E as [4]

$$\nabla^2 E_i = -\omega^2\mu\epsilon E_i = -\frac{\omega^2}{c^2}N^2(x,y)E_i = -k_0^2 N^2(x,y)E_i \tag{4.3}$$

where k_0 is the free space wave vector, c is the velocity of light in free space and $N(x, y)$ denotes the refractive index distribution of the medium. We have assumed that μ is constant in the dielectric waveguides and is equal to that in free space [20]. The subscript i stands for x, y or z, depending on the direction of the component of the vector. A similar wave equation exists for the magnetic field vector H [4], i.e.

$$\nabla^2 H_i = -k_0 N^2(x,y)H_i \tag{4.4}$$

Either equation (4.3) or equation (4.4) can be used to find one of the components of E or H, and the rest of the components of both E and H can be found by equation (4.1). Usually it is preferable to work with equation (4.3) as the electric field is responsible for most physical phenomena in optics [17], and is the principal field measured by photo-detectors [12].

The solution of the wave equation (4.3) is quite complex for rectangular dielectric structures [3]. However, as we are considering a planar slab structure extending to infinity along both the $+x$ and $-x$ directions (Figure 4.1), a very simple form of analytical solution can be obtained [2, 4]. Inspection of the structure shows that either E_z or H_z can exist in this structure, but not both simultaneously (this is because of the absence of any boundary which is perpendicular to the x-direction). Hence by enforcing the continuity condition it can be seen that only the components E_z or H_z exist [11]. This leads to two possible solutions: a *transverse electric* (TE) mode with $E_z = 0$, or a *transverse magnetic* (TM) mode with $H_z = 0$. The notation used follows that which is commonly used in metallic waveguides [2]. Equations (4.1) and (4.2) can be expanded to give the following set of relations between the different components of the electric and magnetic field vectors:

$$\frac{\partial E_z}{\partial y} - \frac{\partial E_y}{\partial z} = -j\omega\mu H_x \tag{4.5}$$

$$\frac{\partial E_z}{\partial x} - \frac{\partial E_x}{\partial z} = j\omega\mu H_y \tag{4.6}$$

$$\frac{\partial E_y}{\partial x} - \frac{\partial E_x}{\partial y} = -j\omega\mu H \tag{4.7}$$

$$\frac{\partial H_z}{\partial y} - \frac{\partial H_y}{\partial z} = j\omega\epsilon E_x \tag{4.8}$$

$$\frac{\partial H_z}{\partial x} - \frac{\partial H_x}{\partial z} = -j\omega\epsilon E_y \tag{4.9}$$

$$\frac{\partial H_y}{\partial x} - \frac{\partial H_x}{\partial y} = j\omega\epsilon E_z \tag{4.10}$$

For the fields to propagate in the TE mode, we require $H_z \neq 0$, $E_x \neq 0$, $E_y = 0$, $H_y \neq 0$ and $H_x = 0$, otherwise the above curl equations cannot be satisfied. Using similar arguments, it can be shown that for the TM mode, we must have $E_x = 0$, $E_y \neq 0$, $H_y = 0$ and $H_x \neq 0$ [3, 4]. It can be seen that the electric field on the transverse x–y plane of the TE mode is orthogonal to that of the TM mode (in fact, we can prove this analytically from the solutions of TE and TM modes tabulated in Table 4.1) [18, 19]. The field solution for the TE mode can be found by substituting E_x into equation (4.3) with $E_x \propto \exp(-j\beta z)$ [2]:

$$\frac{\partial^2 E_x}{\partial y^2} = -(k_0^2 N_1^2 - \beta_z^2)E_x = -q^2 E_x \tag{4.11}$$

in a region with a refractive index N_1 (hereafter denoted as region 1) (Figure 4.1). We have assumed that E_x is independent of x because of the infinite dimensions of the slab (see equations (4.5) and (4.6)). The parameter β_z is known as the *longitudinal propagation constant*. It determines how the phase of the electric field varies with distance z in the waveguide. If the propagation

Table 4.1 Field expressions for both TE and TM modes in a slab waveguide

Mode	Field solution		Eigen equation
TE			
$E_z = E_y = 0;\ E_x = \begin{cases} A\cos(qy) + B\sin(qy), \\ A\exp(-ry), \\ A\cos(qd) - B\sin(qd)\exp[r(y+d)], \end{cases}$ $H_y = H_z = 0$	$\begin{array}{l} -d \leqslant y \leqslant 0 \\ y \geqslant 0 \\ y \leqslant -d \end{array}$		$\tan(dq) = \dfrac{2rq}{q^2 - r^2}$
TM			
$H_z = H_y = 0;\ H_x = \begin{cases} C\cos(qy) + D\sin(qy), \\ C\exp(-ry), \\ C\cos(qd) - D\sin(qd)\exp[r(y+d)], \end{cases}$ $E_z = E_y = 0$	$\begin{array}{l} -d \leqslant y \leqslant 0 \\ y \geqslant 0 \\ y \leqslant -d \end{array}$		$\tan(dq) = \dfrac{2n_1^2 n_2^2 qr}{n_1^4 q^2 - n_1^4 r^2}$

Note: $\beta^2 = k^2 n_1^2 - q^2$

Only the solution of one transverse field component is given. The remaining field solutions can be found by using Maxwell's equations. The arbitrary constants A, B, C and D can be found by the output power condition of the amplifier.

constant is real, then we have travelling fields with the phase varying sinusoidally with z [2]. If β_z is purely imaginary, the fields will vary exponentially with z and are commonly known as *evanescent waves* [11, 17]. In many cases β_z is complex, and the resulting fields are both travelling with their phases varying sinusoidally with z and their amplitudes changing exponentially with z. Let us proceed with the analysis in this dielectric slab (Figure 4.1) by considering the fields outside region 1. In the cladding region (i.e. regions on top of and below region 1), if $N_2 = N_3$, the wave equation becomes

$$\frac{\partial^2 E_x}{\partial y^2} = -(k_0^2 N_2^2 - \beta_z^2)E_x = -r^2 E_x \qquad (4.12)$$

Both the constants q^2 and r^2 can be positive or negative, depending on the relative magnitude of N_1, N_2, k_0 and β_z. The possible solutions for different ranges of negative values of q^2 and r^2 are illustrated in Figure 4.2 [12]. If both q^2 and r^2 are positive (see Figure 4.2), the corresponding values of the propagation constant β_z are in the *continuous radiating mode* regime [20]. The resulting waveguide does not guide any optical signals. For *guided modes* (see the middle curve in Figure 4.2) to occur, q^2 has to be positive but r^2 must be negative, i.e. $k_0^2 N_1^2 > \beta_z^2 > k_0^2 N_2^2$ [21]. In this case the solutions for the transverse fields can be found by imposing continuity requirements at the appropriate boundaries:

$$\begin{array}{ll} E_x = C_1 \exp(-ry), & y \geqslant 0 \\ E_x = C_1 \cos(qy) + C_2 \sin(qY), & -d \leqslant y \leqslant 0 \qquad (4.13) \\ E_x = [C_1 \cos(qd) - C_2 \sin(qd)]\exp[r(y+d)], & y \leqslant -d \end{array}$$

where we have dropped the term $\exp(-j\beta_z z)$, assuming that it is implicit. Equation (4.13) is a general solution. There are four unknowns yet to be

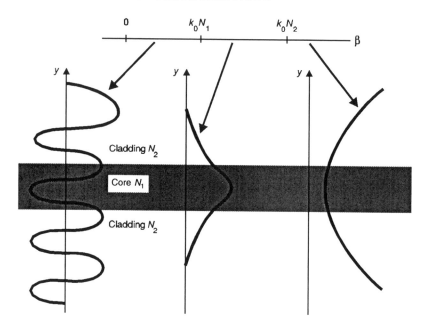

Figure 4.2 Different types of modal solutions determined by the constants q and r

determined: C_1 and C_2, which correspond to the amplitude of the transverse modal field, and q and r, which determine the exact "shape" of the field as well as the value of the propagation constant β_z. To find the values of q and r that satisfy equation (4.13) for a given value of k_0, N_1 and N_2, we have to impose an additional boundary condition for the continuity of H_z at the boundaries. Application of equation (4.7) gives

$$H_z = \frac{1}{j\omega\mu}\frac{\partial E_x}{\partial y} \tag{4.14}$$

Substituting equation (4.13) into equation (4.14) at the appropriate boundaries, we obtain the following relations:

$$-rC_1 = qC_2 \tag{4.15}$$

$$\frac{q}{r}[C_1 \tan(qd) - C_2] = C_1 - C_2 \tan(qd) \tag{4.16}$$

By using equation (4.15) in (4.16), we arrive at the *eigen equation* which gives the values of q and r satisfying the requirements for guided modes:

$$\tan(qd) = \frac{2qr}{q^2 - r^2} \tag{4.17}$$

where d is the thickness of the slab (Figure 4.1). For a guided mode solution, which is our primary interest here, equation (4.17) can be solved by noting that

$$q^2 = k_0^2 N_1^2 - \beta_z^2 \tag{4.18}$$

$$r^2 = \beta_z^2 - k_0^2 N_2^2 = k_0^2(N_1^2 - N_2^2) - q^2 \tag{4.19}$$

By solving equations (4.17) and (4.19) iteratively, we can arrive at solutions of q and r for a particular structure (hence N_1 and N_2) and frequency (hence the wave number k_0). Once the values of q and r have been determined, the unknown C_1 and C_2 can be found easily if the total power of the propagating electromagnetic wave is known. It should be noted that, because of the periodicity of equation (4.17), each particular set of values of k_0, N_1 and N_2 will result in a number of possible values of q and r that will satisfy equation (4.17). We will discuss this implication later. Let us first examine the property of equation (4.17) at $y = -d/2$, i.e. at the plane situated at the centre in region 1. If we denote $\tan(qd/2)$ as t, and by using trigonometric identities, we will arrive at the following relation:

$$\tan(qd) = \frac{2t}{1 - t^2} = \frac{2qr}{q^2 - r^2} = \frac{2(r/q)}{1 - (r/q)^2} \tag{4.20}$$

which, by solving in terms of t, gives [3]

$$t_1 = \tan\left(\frac{qd}{2}\right) = \frac{r}{q} \tag{4.21}$$

$$t_2 = \tan\left(\frac{qd}{2}\right) = -\frac{q}{r} \tag{4.22}$$

Equation (4.21) represents *even* mode solutions, with a maximum in E_x at $y = -d/2$, whilst equation (4.22) represents *odd* mode solutions, with $E_x = 0$ at $y = -d/2$ [3]. The procedures outlined above are for solutions of the TE mode only. The corresponding field solutions for TM modes and the eigen equations for determining the guided mode values of q and r can be found by following this approach, but with H_x replacing E_x. Details of the solution procedure for the TM mode can be found in standard textbooks (e.g. [3, 4]). Here, for the sake of completeness, we will only quote the solutions and eigen equations for the TE and TM modes of the planar dielectric slab waveguide; these are tabulated in Table 4.1.

Before we examine dielectric waveguides with more complex structures, we must first discuss the concept of *modes*. The forms of the eigen equations for q and r of the TE and TM modes require both q and r to be a set of *discrete* values for a particular value of k_0 and a particular refractive index distribution (i.e. N_1 and N_2). These values of q and r are called the *eigenvalues* of the eigen equation. This is illustrated in Figure 4.3, where the functional forms of equations (4.21) and (4.22) are plotted with the constraints of equations (4.18) and (4.19) [12]. A similar figure holds for the eigen equation of the TM mode. The circular arcs are derived from equations (4.18) and (4.19), which have the functional form

$$(qd/2)^2 + (rd/2)^2 = k_0^2 d^2 (N_1^2 + N_2^2)/4 = \text{constant}$$

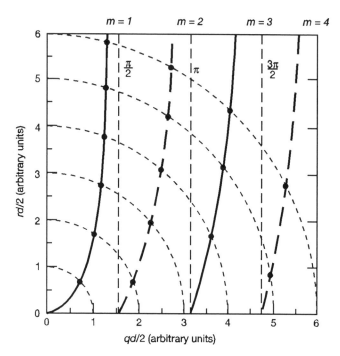

Figure 4.3 Graphical method to solve the eigen equation for a symmetrical three-layer dielectric waveguide (after reference [12])

The intersections of equation (4.21) (solid lines in Figure 4.3) and equation (4.22) (dashed lines in Figure 4.3), with one of those arcs representing the possible solutions for the eigen equation with a particular value of $k_0^2 d^2(N_1^2 + N_2^2)/4$. It then follows from equations (4.18) and (4.19) that hat β_z must also have a finite number of discrete values. Each of these discrete values of β_z is said to be the propagation constant of a *mode* with order m. Notice that the previous description of *even* and *odd* modes does not describe whether the value of m is even or odd. It should be emphasised that, on the contrary, the modal field of an *odd* mode (with a zero value of E_x at $y = -d/2$) has an *even* value of m, whereas the modal field of an *even* mode (with a maxima of E_x at $y = -d/2$) has an *odd* value of m. These different modes will also have distinct transverse modal field distributions, as they have different values of q and r in equation (4.13) [4]. It can also be seen that for some combinations of k_0 (i.e. frequency) and structural parameters d, N_1 and N_2, the number of possible solution modes that satisfy the eigen equation can be as few as one (e.g. the $k_0^2 d^2(N_1^2 + N_2^2)/4 = 1$ has only one intersection with equation (4.21) at $m = 1$). Therefore by using proper structural dimensions, it is possible to design a dielectric waveguide which only supports one single propagating mode *of a particular polarisation* (either the TE or TM mode) for a particular range of signal frequencies. For simple three-layer structures, Figure 4.3 proves to be an extremely useful graphical tool to aid such design. However, for more complex structures like buried heterostructures (BH) in SLAs, the waveguide structure involved is more complicated, and such simple techniques may not be applicable. We will next examine methods to find the transverse modal fields and the corresponding propagation constants in more complex waveguides.

4.2.2 Solution for a rectangular dielectric waveguide using modal field approximations

In theory, the above principles for analysing planar waveguides can be extended to rectangular dielectric structures. A typical rectangular dielectric waveguide is shown in Figure 4.4, with width W and thickness d. To analyse the problem, we can follow the procedures applied to the planar waveguide in the previous section to find the field solutions for rectangular structures as well. We can first generate the wave equation for each region in Figure 4.4, and solve them by matching proper boundary conditions to obtain the appropriate eigen equations and hence the necessary conditions for guided modes. Because of the presence of boundaries (discontinuities in refractive indices) in both the x and y directions, E_z and H_z will be finite for all propagating modes [3]. This will make the analysis more complex, although mathematically feasible.

However, there is an additional problem. The matching of boundary conditions can only be made by choosing either the continuous tangential fields at the four boundaries surrounding region 1 (the *core*) or the normal fields at the same boundaries [22]. This leaves other boundaries (e.g. between regions 3 and 5, 2 and 7, etc.) unmatched. In other words, the boundary conditions cannot be matched simultaneously at all boundaries [4]. Because of this, it is obvious that the field solutions are no longer separable functions of x and y [16, 23–25]. Marcatilli [26] suggested that if the waveguide is supporting a propagating mode which is well above cut-off, then the power flow in the corners (the shaded regions) in Figure 4.4 can be ignored. Then the matching of boundary conditions in these regions is unimportant. Hence we can simply match boundary conditions around the core to generate approximate modal field solutions for the unshaded regions 1–5. In this way, Marcatilli was able to

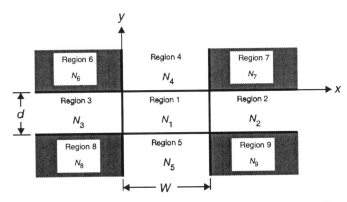

Figure 4.4   The structure used to analyse the modal field distributions in rectangular waveguides

derive analytical expressions, separable in x and y, to approximate the modal field distributions, and developed the corresponding eigen equations for guided mode conditions [3]. He found that the modal field solutions of these eigen equations for the rectangular dielectric waveguide, like the TE and TM modes in a planar dielectric waveguide discussed in subsection 4.2.1, can also be classified into two orthogonal polarisations. The one for which the electric field is strongly polarised along the x-direction is known as the E_{pq}^x mode. It is also known as the *quasi-TE* mode for simplicity, as it is similar to a TE mode *except* for the presence of *both* finite longitudinal components of the electric E_z and the magnetic field H_z [27]. Another one, for which the electric field is strongly polarised along the y-direction, is known as the E_{pq}^y mode, or *quasi-TM* mode. The mode numbers p and q denote the number of zeros along the x- and y-directions, respectively [26]. Notice that the quantity q here is similar in meaning to that used in the previous subsection, where it was used to denote the number of zeros along y in a dielectric slab. The above modal fields in a rectangular dielectric waveguide are known as *hybrid modes* because they are mixtures of the TE and TM modes in the simple three-layer dielectric waveguide discussed in the previous subsection [2]. We have tabulated the analytical solutions for the approximate modal fields in regions 1–5 in Table 4.2 for the field components E_z and H_z, together with the eigen equations derived for the guided modes. Other field components can be found from equations (4.5)–(4.10). The two parameters k_x and k_y in Table 4.2 are known as the *transverse wave numbers* of the modal fields, and are related to the propagation constant β_z by

$$\beta_z^2 = k_0^2 N_1^2 - k_x^2 - k_y^2 \qquad (4.23)$$

Note the similarity between equations (4.23) and (4.18). These two parameters determine the sinusoidal variation form of the fields along the x- and y-directions in the core. We can also relate the *decaying parameters* γ_j in the ith region of the cladding with β_z [4]:

$$\beta_z^2 = k_0^2 N_1^2 - k_y^2 + \gamma_i^2, \quad i = 2, 3 \qquad (4.24)$$

$$\beta_z^2 = k_0^2 N_1^2 - k_x^2 + \gamma_j^2, \quad j = 4, 5 \qquad (4.25)$$

Again, γ_i, k_x and k_y are discrete, which can be found by the eigen equations in Table 4.2 for guided modes. Therefore, similar to the arguments outlined in the previous subsection with Figure 4.3, each modal field solution has a particular value of β_z, and for a given value of the frequency there can be several possible modal field distributions as long as the structure satisfies the eigen equations for multiple guided modes. By manipulating the structural parameters of the waveguide, it is possible to reduce the number of possible modes that can propagate *for a particular polarisation* (i.e. quasi-TE or quasi-TM) to one. Furthermore, it is possible to obtain a structure such that the transverse field is cut off along the x-direction (i.e. $k_y = 0$). This condition is known as the *single transverse mode* (STM) condition [28]. Under such a condition the rectangular waveguide structure can be approximated accurately by the three-layer slab

Table 4.2   Field solutions of wave equations in a rectangular dielectric waveguide

Field Region		E_z	H_z
E^x_{mn}	1	$A\cos(k_x(x+\zeta))\cos(k_y(y+\eta))$	$A\sqrt{\epsilon_0/\mu_0}\,n_1^2(k_y/k_x)(k/\beta)\sin(k_x(x+\zeta))\sin(k_y(y+\eta))$
	2	$A\cos(k_x(W+\zeta))\cos(k_y(y+\eta))\exp[-\gamma_2(x-W)]$	$-A\sqrt{\epsilon_0/\mu_0}\,n_1^2(k_y/\gamma_2)(k/\beta)\cos(k_x(W+\zeta))\sin(k_y(y+\eta))\exp[-\gamma_2(x-W)]$
	3	$A\cos(k_x\zeta)\cos(k_y(y+\eta))\exp(\gamma_3 x)$	$A\sqrt{\epsilon_0/\mu_0}\,n_3^2(k_y/\gamma_3)(k/\beta)\cos(k_x\zeta)\sin(k_y(y+\eta))\exp(\gamma_3 x)$
	4	$A(n_1/n_4)^2\cos(k_x(x+\zeta))\cos(k_y\eta)\exp(-\gamma_4 y)$	$A\sqrt{\epsilon_0/\mu_0}\,n_1^2(\gamma_4/k_x)(k/\beta)\sin(k_x(x+\zeta))\cos(k_y\eta)\exp(-\gamma_4 y)$
	5	$A(n_1/n_5)^2\cos(k_x(x+\zeta))\cos(k_y(\eta-d))\exp[\gamma_5(y+d)]$	$-A\sqrt{\epsilon_0/\mu_0}\,n_1^2(\gamma_5/k_x)(k/\beta)\sin(k_x(x+\zeta))\cos(k_y(\eta-d))\exp[\gamma_5(y+d)]$
E^y_{mn}	1	$A\cos(k_x(x+\zeta))\cos(k_y(y+\eta))$	$-A\sqrt{\epsilon_0/\mu_0}\,n_1^2(k_x/k_y)(k/\beta)\sin(k_x(x+\zeta))\sin(k_y(y+\eta))$
	2	$A(n_1/n_2)^2\cos(k_x(W+\zeta))\cos(k_y(y+\eta))\exp[\gamma_2(x-W)]$	$-A\sqrt{\epsilon_0/\mu_0}\,n_1^2(\gamma_2/k_y)(k/\beta)\cos(k_x(W+\zeta))\sin(k_y(y+\eta))\exp[-\gamma_2(x-W)]$
	3	$A(n_1/n_3)^2\cos(k_x\zeta)\cos(k_y(y+\eta))\exp(\gamma_3 x)$	$A\sqrt{\epsilon_0/\mu_0}\,n_1^2(\gamma_3/k_y)(k/\beta)\cos(k_x\zeta)\sin(k_y(y+\eta))\exp(\gamma_3 x)$
	4	$A\cos(k_x(x+\zeta))\cos(k_y\eta)\exp(-\gamma_4 y)$	$A\sqrt{\epsilon_0/\mu_0}\,n_4^2(k_x/\gamma_4)(k/\beta)\sin(k_x(x+\zeta))\cos(k_y\eta)\exp(-\gamma_4 y)$
	5	$A\cos(k_x(x+\zeta))\cos(k_y(\eta-d))\exp[\gamma_5(y+d)]$	$-A\sqrt{\epsilon_0/\mu_0}\,n_5^2(k_x/\gamma_5)(k/\beta)\sin(k_x(x+\zeta))\cos(k_y(\eta-d))\exp[\gamma_5(y-d)]$

For the E^x_{mn} mode:

$$\tan(k_y d) = \frac{k_y(\gamma_4+\gamma_5)}{k_y^2-\gamma_4\gamma_5}, \qquad \beta^2 = k_0^2 n_1^2 - k_x^2 - k_y^2$$

$$\tan(k_x W) = \frac{n_1^2 k_x(n_2^2\gamma_3+n_3^2\gamma_2)}{n_2^2 n_3^2 k_x^2 - n_1^4\gamma_2\gamma_3}$$

$$\tan(k_x\zeta) = \left(\frac{n_3}{n_1}\right)^2\left(\frac{k_x}{\gamma_3}\right), \qquad \tan(k_y\eta) = \frac{\gamma_4}{k_y}$$

For the E^y_{mn} mode:

$$\tan(k_x W) = \frac{k_x(\gamma_2+\gamma_3)}{k_x^2-\gamma_2\gamma_3}, \qquad \beta^2 = k_0^2 n_1^2 - k_x^2 - k_y^2$$

$$\tan(k_y d) = \frac{n_1^2 k_y(n_4^2\gamma_5+n_5^2\gamma_4)}{n_4^2 n_5^2 k_y^2 - n_1^4\gamma_4\gamma_5}$$

$$\tan(k_y\eta) = -\left(\frac{n_4}{n_1}\right)^2\left(\frac{k_y}{\gamma_4}\right), \qquad \tan(k_x\zeta) = -\frac{\gamma_3}{k_x}$$

discussed in the previous subsection [5, 29]. Marcatilli's work provided a set of convenient analytical solutions to approximate the modal field distributions in a rectangular dielectric waveguide. However, the assumption of these functions being separable in x and y is not valid near cut-off, i.e. when the power flow in the corner regions cannot be ignored [13, 30]. For a better range of solutions, alternative numerical techniques have to be used. A convenient technique known as the *effective index method* (EIM) provides a simple method to obtain more accurate values of β_z for a wider range of frequencies, and this will be discussed below [4, 31, 32].

4.2.3 Application of EIM for calculating propagation constants for transverse modal fields in rectangular dielectric waveguides

In order to improve the solution range of β_z, an alternative method is to intercouple the modal field approximations in the x- and y-directions using a concept known as the *effective index*. The accuracy of β_z can be improved by taking into account the interrelation between the transverse wave numbers k_x and k_y, which were assumed to be independent of each other in subsection 4.2.2. Physically, this accounts for the power flow in the active region more accurately than modal field approximations. The effective index for the mth propagating mode can be found by [1]

$$N_{em}^2 = \left(\frac{\beta_z}{k_0}\right)^2 = \frac{\iint N^2(x,y)|\Phi_m(x,y)|^2 \mathrm{d}x\mathrm{d}y}{\iint |\Phi_m(x,y)|^2 \mathrm{d}x\mathrm{d}y} \tag{4.26}$$

where Φ_m is the mth transverse modal field solution of the wave equation for the structure, and integration is taken over the entire transverse x–y plane. The index m indicates a particular combination of number of zeros along x and y (i.e. p and q), respectively. This is particularly useful in actual programming to solve for β_z in computers (see, for example, the algorithms proposed by Henry *et al.* [33, 34]). It can be seen from the definition in equation (4.26) that N_{em} is a *group* index because it determines the group velocity of the propagating fields [2]. Suppose that $\Phi_m(x, y)$ is a separable function in x and y, such that $\Phi_m(x,y) = F(x)G(y)$. We can now find the effective index N_{ey} of the dielectric slab of indices N_4, N_1 and N_5 with thickness d for the propagating mode (see Figure 4.5(a)). For the quasi-TE mode, N_{ey} can be found by solving the eigen equation for the dielectric slab along y. This equation is given by [3]

$$\tan(k_y d) = \frac{k_y(\gamma_4 + \gamma_5)}{k_y^2 - \gamma_4 \gamma_5} \tag{4.27}$$

This formulation is possible because $\Phi_m(x, y)$ is a separable function and hence eigen equations can still be used [2, 25]. The parameters γ_4 and γ_5 are given by

$$\gamma_j^2 = k_0^2(N_1^2 - N_j^2) - k_y^2, \quad j = 4, 5 \tag{4.28}$$

Once k_y is found, N_{ey} can be found by [4]

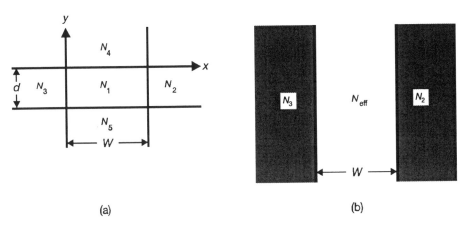

Figure 4.5 Solution procedure for analysing a rectangular waveguide using the EIM

$$N_{ey}^2 = N_1^2 - \left(\frac{k_y}{k_0}\right)^2 \qquad (4.29)$$

We then solve for the dielectric slab in the x-direction with indices N_3, N_1 and N_2, as shown in Figure 4.5(b). For the quasi-TE mode, the eigen equation will be [3]

$$\tan(k_x W) = \frac{N_{ey}^2 k_x (N_2^2 \gamma_3 + N_3^2 \gamma_2)}{N_2^2 N_3^2 k_x^2 - N_1^4 \gamma_2 \gamma_3} \qquad (4.30)$$

and the decaying parameters γ_2 and γ_3 are given by [4]

$$\gamma_i^2 = k^2 (N_{ey}^2 - N_i^2) - k_x^2, \quad i = 2, 3 \qquad (4.31)$$

The effective index of the entire waveguide is given by

$$N_{em}^2 = N_{ey}^2 - \left(\frac{k_x}{k_0}\right)^2 = N_1^2 - \left(\frac{k_x}{k_0}\right)^2 - \left(\frac{k_y}{k_0}\right)^2 \qquad (4.32)$$

Note that equation (4.32) is an equivalent form of equation (4.23). From this equation, N_{em} can be calculated and β_z can subsequently be found from equation (4.26). The procedures outlined above are for solutions of quasi-TE modes only. A similar procedure is adopted for solutions of quasi-TM modes, except with different eigen equations. To find N_{ey}, the following eigen equation for the quasi-TM mode (see Table 4.2) first has to be solved:

$$\tan(k_y d) = \frac{N_1^2 k_y (N_4^2 \gamma_5 + N_5^2 \gamma_4)}{N_4^2 N_5^2 k_y^2 - N_1^4 \gamma_4 \gamma_5} \qquad (4.33)$$

followed by equation (4.29). The decaying parameters γ_4 and γ_5 can be found from equation (4.28). Then the effective index N_{em} can be found by first solving for k_x by

$$\tan(k_x W) = \frac{k_4(\gamma_2 + \gamma_3)}{k_x^2 - \gamma_2 \gamma_3} \tag{4.34}$$

where γ_2 and γ_3 can be found from equation (4.31), and finally equation (4.32) can be used to calculate N_{em}. The major differences between Marcatilli's modal approximation and EIM can be seen by comparing equations (4.24) and (4.25) with equation (4.31). Marcatilli assumed that the decaying parameters can be calculated from the same refractive index of the active region N_1, whereas with EIM we have taken into account the energy leaked to the cladding of N_4 and N_5 before calculating γ_2 and γ_3 by using equation (4.31). Physically speaking, the inter-coupling effect between the decaying parameters and the transverse wave numbers has been accounted for, and hence a more accurate description of the power flow in the structure can be obtained. Consequently, the EIM generates a value of β_z that is more accurate than Marcatilli's modal field approximation technique when compared with more exact solutions obtained by rigorous numerical analysis, as well as with experimental measurements [4, 35].

There are two major disadvantages in using EIM to solve the wave equations in rectangular dielectric waveguides. First, EIM does not generate field solutions $\Phi_m(x, y)$ automatically [31]. Additional numerical techniques have to be used if equation (4.26) has to be used to derive $\Phi_m(x, y)$ from the effective index N_{em} (for example, see the complex field expressions used by Buus [6] in analysing gain-guided semiconductor lasers). In addition, the value of β_z calculated from EIM, although more accurate than that obtained by the approximate modal field analyses proposed by Marcatilli, will not be an exact solution [30]. β_z is only exact when a variational analysis performed by equation (4.26) indicates that it is minimal [1, 14, 18]. In EIM, we have assumed that $\Phi_m(x, y)$ is a separable function in x and y, and we have used approximate eigen equations developed by Marcatilli. Hence the same assumption on negligible power flow in the shaded regions in Figure 4.4 still apply to EIM, and a variational analysis by equation (4.26) with the value of β_z obtained by EIM indicates that the expression is not yet minimal, i.e. the corresponding values of β_z do not give an exact solution. These two major disadvantages will create problems for more complex waveguide structures (e.g. rib waveguides [23]), but since many semiconductor laser amplifiers employ relatively simple rectangular buried heterostructures with index guiding (i.e. the fields are real [4–6, 29, 30]), the value of β_z calculated by EIM is sufficiently accurate for most purposes [35], and the modal fields solutions listed in Table 4.2 can be used to a good approximation to describe the fields *well above cut-off*. Hence, in the foregoing analyses and discussions, the propagation constant β_z will be calculated using EIM, and the field solutions will be those proposed by Marcatilli, listed in Table 4.2.

Dispersion curves, i.e. the variation of the propagation constant β_z against the frequency of the propagating fields, have been plotted in Figure 4.6 for a rectangular dielectric waveguide with an aspect ratio $W/d = 2$ and $N_2 = N_3 = N_4 = N_5$ (i.e. a symmetrical waveguide). Because the value of β_z depends on the structure of the waveguide as well as the frequency of the

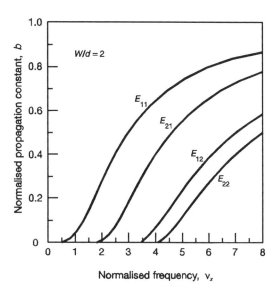

Figure 4.6 Dispersion of curves of b against ν_x for a weakly guiding rectangular waveguide with an aspect ratio $W/d = 2$ calculated by the effective index method

propagating fields, these dispersion curves are conventionally plotted with the *normalised* propagation constant b against the *normalised* frequency ν_x [1, 35, 37, 38]. These parameters are defined by [4]

$$\nu_x^2 = W^2 k_0(N_1^2 - N_2^2) \tag{4.35}$$

$$b^2 = \frac{(\beta_z^2 - k_0^2 N_2^2)}{(k_0^2 N_1^2 - k_0^2 N_2^2)} \tag{4.36}$$

The four lowest order modes are shown in Figure 4.6. In the analysis we have assumed that the waveguide is a *weakly guiding structure* (i.e. $N_1 - N_2 \ll N_1$). Under such circumstances the quasi-TE and quasi-TM modes have virtually identical values of the propagation constant [1, 4]. It can be seen that for all modes, b increases with increasing ν_x. This implies that the propagation constant β_z increases with increasing wave number k_0 and hence increasing signal frequency. Physically, as observed from equation (4.26), an increase in β_z means that more optical power will be confined within the core region (with refraction index N_1) of the rectangular waveguide (see Figure 4.4).

In Figure 4.7 we have compared the dispersion curves for the lowest order modes for several weakly guiding dielectric rectangular waveguides with the same refraction index distribution but different aspect ratios. It can be seen that the value of b changes with the value of the aspect ratio W/d for a particular value of ν_x. The above observation from these dispersion curves is important in designing the dimensions of a rectangular dielectric waveguide. Moreover, these curves also provide important information for the analysis of semiconductor lasers and the optical gain in semiconductor laser amplifiers

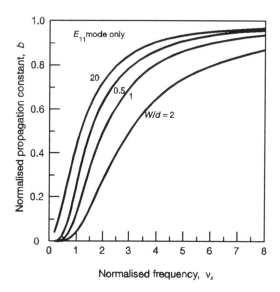

Figure 4.7 Comparisons of dispersion curves for rectangular dielectric waveguides with different aspect ratios

(which have structures that resemble that of a dielectric rectangular waveguide). An example of the former analysis was performed by Buus [6, 39] who used the EIM to evaluate the threshold current required for different semiconductor laser diode structures.

4.2.4 Other methods to solve for transverse modal fields and the dispersion characteristics of rectangular dielectric waveguides

Marcatilli's approximations and the EIM form the basis for analysing rectangular dielectric waveguides. They also form the basis for a number of variant methods. A recent one can be found in Cheung and Lin's work [31] on the iterative equivalent index method and Chiang's [32] modified EIM with perturbation corrections. For more accurate solutions, other methods of solving the wave equations have to be used. For applications using all of these methods, accuracy and computational efficiency are the most important factors, and it is obvious that they are often in conflict. The most reliable and accurate method can be found in Goell's [38] landmark work using point-matching with Bessel functions. Indeed, it is so accurate that other numerical techniques have to compare with it to verify their accuracy [3, 4]. With powerful computers, Goell's approach can be implemented easily. A recent attempt to solve by exact solution using an alternative series of basis functions with the concepts of *residues* can be found in reference [40]. Another rigorous method is to employ the *finite element method* (FEM) [41]. The benefit of using the FEM is that vectorial wave equations can also be solved, giving polarisation corrections due to the anisotropic characteristics of the material [16].

Another approach which has been reported is the integral equation approach [18]. This technique is well known in analysing other scattering problems [42, 43]. It is assumed that the fields propagating in the *core* region of the waveguide are excited by equivalent current sources. In this way, a domain integral equation can be formulated [15, 44]. This formulation is very general and can be used for many different waveguide structures with many stratified layers. It also works for vectorial wave equations. However, solving the integral equations requires the use of a very powerful computer, usually in the supercomputer regime, and is also very time consuming. Some faster methods based on modification of integral equations can be found in Hoekstra's work [45].

A more economical way is to use variational analysis on the expression of equation (4.26). By iteratively substituting approximations of $\Phi_m(x, y)$ into it, we can obtain the exact solution of β_z as well as the modal field solutions $\Phi_m(x, y)$ by minimising the expression [4]. There are different ways to find the minimum of equation (4.26). It can be based upon the expansion of a series of basic functions and the iterative deduction of coefficients for each term in the series until β_z is a minimum [4]. Alternatively, direct iteration by combining equation (4.26) with the perturbation expression (see the next section) can also be used [30]. A third way is to start from a separable function for $\Phi_m(x, y)$ and then proceed to find the solutions of $F(x)$ and $G(y)$ by weighting the refractive indices, i.e. by using the method of moments [2]. This method, known as the *weighted index method* (WIM), is extremely fast, accurate as well as highly efficient, and can yield accurate solutions even for rib waveguides [23, 24]. Other methods of course exist: e.g. the steepest descent approximation technique to solve the wave equation [46]; the vector beam propagation method [30]; or a two-dimensional Fourier analysis [33, 34].

4.3 APPLICATIONS OF SOLUTIONS OF TRANSVERSE MODAL FIELDS IN SLAs

We saw in Chapter 3 that in analysing the gain characteristics of SLAs, one of the crucial tasks is to relate the material gain coefficient g_m to the single pass gain G_s. In general, if the losses in the cladding can be ignored, it is given by

$$G_s = \exp[\Gamma_m(g_m - \alpha_a)L] \qquad (4.37)$$

where α_a is the loss in the active region of the amplifier, and Γ_m is the optical confinement factor of the mth propagating mode. We can replace the exponent in this equation by the *modal gain coefficient g* [47], such that

$$G_s = \exp(gL) \qquad (4.38)$$

Buus [6] called this parameter the *effective gain*. For cases of negligible cladding loss, g is simply given by

$$g = \Gamma_m(g_m - \alpha_a) \qquad (4.39)$$

In the following subsections we examine how to calculate g in more complicated cases with finite cladding losses, and how the structure of the semiconductor laser amplifier (SLA) will affect the value of g, and hence G.

4.3.1 Analysis of the modal gain coefficients

We stated in Chapter 3 that when the loss in the cladding region is finite and uniform, the optical confinement factor Γ_m can be used to calculate the modal gain coefficient g. As shown by Adams [4], this is given by

$$g = \Gamma_m(g_m - \alpha_a) - (1 - \Gamma_m)\alpha_c \qquad (4.40)$$

where α_c is the loss in the cladding region. The above expression can be derived from the variational expression of equation (4.26) [47]. Assuming that, because of the presence of gain in the active region there is an amplitude variation along z as the field is being amplified as it propagates across the amplifier. This amplitude variation can be described by introducing a complex longitudinal propagation constant β_z (cf. the complex propagation constant in transmission lines with losses [2]):

$$\beta_z = \beta_{zr} + j\beta_{zi} \qquad (4.41)$$

Since the total propagating field solution of the wave equation is given by the product of the transverse modal field solution $\phi_m(x, y)$ and $\exp(-j\beta_z z)$ (see section 4.2), we can immediately see that the imaginary part β_{zi} takes into account the variation of the amplitude of the field due to optical gain, which is proportional to $\exp(\beta_{zi}L)$ over an amplifier of length L. In this case, the power amplification factor, i.e. the single pass gain, is given by the square of the field amplitude, i.e.

$$G_s = \exp(-2\beta_{zi}L) \qquad (4.42)$$

Equating equation (3.38) with equation (3.42) gives the following relation:

$$g = -2\beta_{zi} \qquad (4.43)$$

To proceed further, we consider the structure sketched in Figure 4.8, which represents the cross-section of a buried heterostructure (BH) SLA [7]. We compare the wave equations for a passive waveguide shown in Figure 4.4 and that with gain and losses (Figure 4.8). They are listed below:

$$[\nabla_t^2 + k_0^2 \hat{N}^2(x, y) - \hat{\beta}_z^2]\hat{\Phi} = 0 \qquad (4.44)$$

$$[\nabla_t^2 + k_0^2 N^2(x, y) - \beta_z^2]\Phi = 0 \qquad (4.45)$$

where the quantities with a caret ($\char94$) are those for the passive waveguide, assuming that the transverse field solutions are purely real (i.e. are index guided structures) [4, 5]. The subscript m for the transverse modal field solutions has been dropped for simplicity, assuming their dependence on m to be implicit. Multiplying equations (4.44) and (4.45) by Φ and $\hat{\Phi}$, respectively, followed by subtraction from each other, we obtain the following expression under a

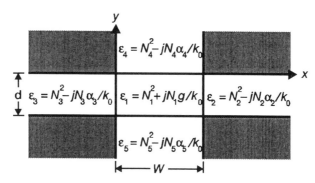

Figure 4.8 The structure used in perturbation analysis of the modal gain coefficients in BH SLAs

weakly guiding approximation after integrating over the entire cross-section S [1]:

$$\beta_z^2 - \hat{\beta}_z^2 = \frac{\iint_s k_0^2 [\epsilon(x, y) - \hat{\epsilon}(x, y)] \Phi \cdot \hat{\Phi} \cdot dS}{\iint_s \Phi \cdot \hat{\Phi} \cdot dS} \tag{4.46}$$

The dielectric constants ϵ for the active structure for each region are shown in Figure 4.8 [48]. It can be shown that the imaginary parts of the dielectric constants due to gain and losses have much smaller magnitudes compared with the real part N_j^2 [6]. Therefore Figure 4.8 can be regarded as a slightly perturbed structure of that in Figure 4.4. Under these conditions the solutions Φ and $\hat{\Phi}$ are very close to each other, and it can be shown that [7]

$$\beta_z^2 - \hat{\beta}_z^2 \approx \frac{\iint_s k_0^2 [\epsilon(x, y) - \hat{\epsilon}(x, y)] \cdot \hat{\Phi}^2 dS}{\iint_s \hat{\Phi}^2 dS} \tag{4.47}$$

The difference in the dielectric constants between the perturbed and unperturbed structures $[\epsilon(x, y) - \hat{\epsilon}(x, y)]$ is simply the imaginary parts $jN_i\chi_i/k$, where χ_i is either the net gain coefficient $g_m - \alpha_a$ in the active region or the losses in the cladding $\alpha_{i(i=2, 3, 4 \text{ and } 5)}$. Hence, if $\hat{\beta}_z = \beta_0$, then

$$\beta_z^2 = \beta_0^2 + j\Lambda \tag{4.48}$$

and the quantity Λ can be written as [7]

$$\Lambda = \frac{k_0^2 \iint_s (N_i\chi_i/k_0) \cdot \hat{\Phi}^2 \cdot dS}{\iint_s \hat{\Phi}^s dS} \tag{4.49}$$

Now since the imaginary parts of the dielectric constants of the perturbed structure of Figure 4.8 are much smaller than their corresponding real parts, the quantity Λ will also be small compared with β_0^2. Applying the binomial theorem and ignoring terms with orders higher than 2, the difference between the propagation constants of a perturbed active waveguide with that of the unperturbed passive waveguide is found to be given by

$$\Delta\beta_z = \beta_z - \beta_0 = \frac{j\Lambda}{2\beta_0} \tag{4.50}$$

Therefore perturbation analysis shows that the introduction of gain and losses into a dielectric waveguide (e.g. in a SLA) actually introduces a small imaginary part to the original propagation constant of an identical unperturbed lossless waveguide [49]. Substituting equation (4.41) into equation (4.50) results in

$$\beta_{zi} = \Delta\beta_z = \frac{j\Lambda}{2\beta_0} \tag{4.51}$$

and

$$\beta_{zr} = \beta_0 \tag{4.52}$$

The modal gain coefficient is obtained by substituting the above equations into equation (4.42). That is

$$g = -2\Delta\beta_z = -\frac{j\Lambda}{\beta_0} \tag{4.53}$$

Therefore if the modal field solutions of the passive (unperturbed) waveguide are known, then the quantity Λ can be calculated easily from equation (4.49), and the modal gain coefficients can then be found from equation (4.53). In general, the quantity Λ for a general perturbed structure with dielectric constant distribution $\epsilon(x, y)$ is given by the following expression:

$$\Lambda = \frac{k_0^2 \iint_s [\epsilon(x, y) - \hat{\epsilon}(x, y)] \hat{\Phi}^2 \cdot \mathrm{d}S}{\iint_s \hat{\Phi}^2 \cdot \mathrm{d}S} \tag{4.54}$$

where $\hat{\Phi}$ is the unperturbed modal field solutions found in the unperturbed structure. For a lossy structure, Λ is positive and g is negative. However, if sufficient material gain is introduced in region 1 (see Figure 4.8), such that Λ becomes negative, then g and hence G_s become positive, and the resulting structure is capable of optical amplification.

The advantages of using perturbation analysis can be seen when the loss distribution in the cladding is not uniform around regions 2–5. In this situation, the expression of equation (4.40) for modal gain coefficients is no longer valid [49]. This is because the power flow in the cladding can no longer be represented by a simple factor $(1 - \Gamma_m)$. Instead, we have to resort to Poynting vectors to calculate the proportion of optical power flow in each region, and then evaluate the overall modal gain coefficient by weighting the losses and gains with these proportions [4, 7]. Furthermore, if the loss and gain coefficients vary in the functional form of x–y (e.g. in a medium with a parabolic gain profile [6]), then it is impossible to use equation (4.40). In this situation the perturbation analysis offers a convenient way to solve the problem numerically using equations (4.54) and (4.53). Because many numerical algorithms are available to solve for the modal field solutions of passive optical waveguides, the solutions can be directly input into equation (4.54) to compute the modal gain coefficient g rapidly. In the above derivation we have assumed an index guided structure, such that the transverse field solutions are purely real [1]. However, there exist gain-guided structures in which the gain and loss distributions account for the guiding action [4, 29]. In these structures the modal field solutions are generally complex [36], and have expressions similar to the Gaussian beam description of the propagation of light [12]. In these circumstances the above expressions are still applicable, but with the square of the field $\hat{\Phi}^2$ replaced by the product of the complex field with its conjugate $\hat{\Phi} \cdot \hat{\Phi}^*$. The study of these gain-guided laser amplifiers is very challenging and is beyond the scope of this book.

4.3.2 *Design of a polarisation insensitive Travelling Wave Amplifier (TWA)*

In most of the relevant literature the analyses of BH laser diodes and laser amplifiers have been performed under the assumption of single transverse mode (STM) operation, in which the fields are assumed to be cut off along the x-direction (Figure 4.9) [9, 28]. In this case the propagating fields are identical to those propagating in a planar dielectric slab waveguide, which we analysed in subsection 4.2.1. This is in fact necessary in designing a laser diode in order to avoid multi-mode emissions [5]. Saitoh and Mukai [9] analysed the polarisation dependence of signal gain in a travelling-wave amplifier (TWA) for the TE and TM modes under STM operation. They discovered that the difference between the values of the optical confinement factors for these two polarisations, given by equations (4.38) and (4.40), could be minimised by simply increasing the active layer thickness, d. From these studies they have derived design curves for polarisation insensitive TWAs.

However, when the active layer thickness d is increased, depending on the value of W it is possible for the TWA to depart from STM operation by exciting higher order quasi-TE and quasi-TM modes, as shown by Ghafouri-Shiraz [28] in studies on STM conditions for separate confinement hetero-structure (SCH) semiconductor lasers. Therefore, for thick active layers, the polarisation sensitivities of the signal gain of hybrid modes must be considered as well to get a more complete picture before analysing other amplifier characteristics. This can be done easily by analysing the modal gain coefficients for different hybrid modes in addition to those of the TE and TM modes.

We first compared the accuracy of using the perturbation approximation of equations (4.49)–(4.53) to derive the modal gain coefficients with the exact methods calculated using Poynting vectors [4]. In Figure 4.10(a) we have calculated the optical confinement factor Γ_m from the modal gain coefficients

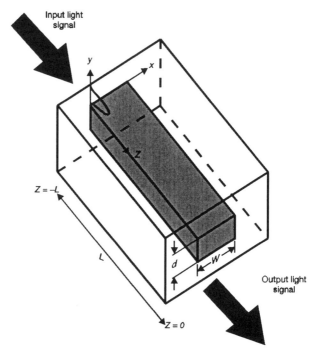

Figure 4.9 Illustration of the form of an electric field in a rectangular dielectric optical waveguide, with a single transverse mode condition (field variation independent of x)

derived by the above two methods using equation (4.40). The parameters used in the analysis are tabulated in Table 4.3. It can be seen that the perturbation approximation yields results in close agreement with those obtained by Poynting vector analysis. This proves the usefulness and accuracy of the perturbation approximation proposed by the author. The difference of taking into account the active layer width W (i.e. using hybrid modes and assuming that the fields are not cut off along x [49]) in the analysis of the optical confinement factor Γ_m is illustrated by the curves in Figure 4.10(b). Here, the optical confinement factor of the fundamental TE and TM modes is plotted against the active layer thickness d from 0 μm to 1 μm. Similar plots for optical

Table 4.3 Parameters used to calculate Figure 4.10(a) and (b)

Active layer width:	$W = 5 \, \mu$m
Active layer thickness:	$d = 0\text{-}1 \, \mu$m
Refractive indices:	$N_1 = 3.524$
	$N_2 = N_3 = N_4 = N_5 = 3.169$
Wavelength of propagating fields:	$\lambda = 1.55 \, \mu$m
Net material gain coefficient:	$(g - \alpha_a) = 100 \, \text{cm}^{-1}$
Loss in cladding:	$\alpha_2 = \alpha_3 = \alpha_4 = \alpha_5 = 50 \, \text{cm}^{-1}$

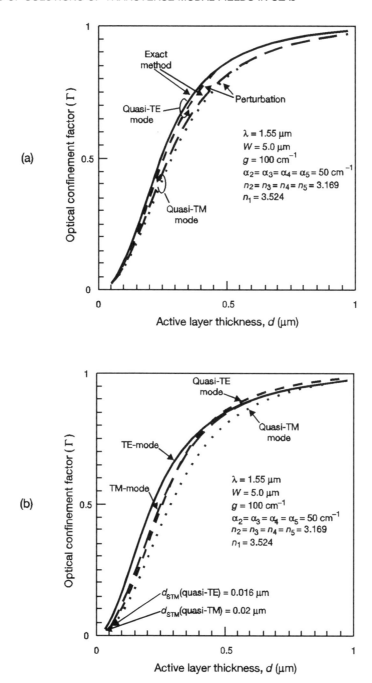

Figure 4.10 (a) Comparison of calculated optical confinement factors for hybrid modes calculated from Poynting vectors and the perturbation approximation, respectively. (b) Variation of optical confinement factor with active layer thickness for the TE and TM modes, and the quasi-TE and quasi-TM modes

confinement factors of the fundamental quasi-TE and quasi-TM modes are also plotted, with the field solutions Φ_m taken from Marcatilli's approximations listed in Table 4.2. We first calculate the modal gain coefficient by the perturbation analysis outlined in the previous section, and derive Γ_m from equation (4.40). As observed in Figure 4.10(b), there are significant differences between the optical confinement factors for slab modes and those for hybrid modes, although both types of field exhibit a reducing polarisation sensitivity with increasing d. We have also indicated the upper limit on d for STM operation in Figure 4.10(b). Obviously, if we attempt to improve the polarisation sensitivity of the TWA by increasing d, the amplifier structure will no longer be operating in the STM, and it is necessary to consider higher order modes like quasi-TE and quasi-TM modes instead of TE and TM modes only. In the following analysis we will take into account the active layer width W in order to get a deeper insight into the structural effects on the polarisation sensitivity of TWAs [49]. In particular, we will consider three slightly different structures with different refractive index distributions.

Effect of active layer thickness

We have calculated the single pass gain G_s for the quasi-TE and quasi-TM modes of a TWA using the perturbation method discussed in subsection 4.3.1 together with equations (4.40) and (4.38). The variation of G_s with active layer thickness d for three different structures is shown in Figures 4.11, 4.13 and 4.15.

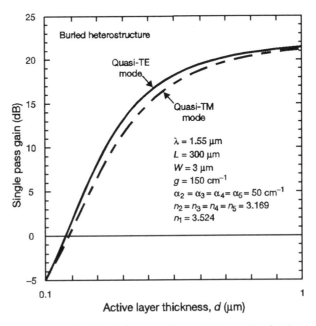

Figure 4.11 Variation of gain with active layer thickness for fundamental quasi-TE and quasi-TM modes in a buried heterostructure TWA

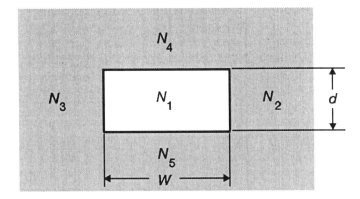

Figure 4.12   Refractive index distribution for a buried heterostructure TWA

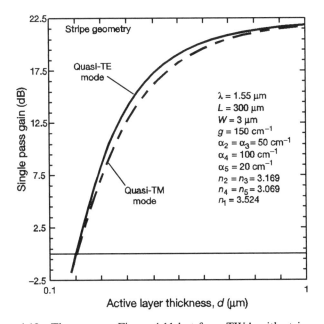

Figure 4.13   The same as Figure 4.11 but for a TWA with stripe geometry

These structures were a buried heterostructure, a stripe geometry structure and a buried channel (BC) structure, respectively. We have illustrated their differences in Figures 4.12, 4.14 and 4.16, and the parameters used in the analysis are tabulated in Table 4.4(a)–(c).

In Figure 4.11 we can see that an increase in the active layer thickness d improves the confinement of both hybrid modes, and hence the value of G_s increases with d for both polarisations. The confinement for the quasi-TM mode is improving faster than that of the quasi-TE mode as the field of the former polarisation is strongly polarised along the thickness of the active

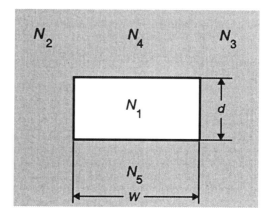

Figure 4.14   Refractive index distribution for a TWA with stripe geometry

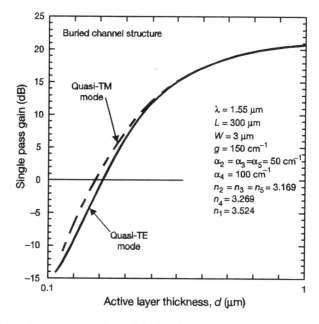

Figure 4.15   The same as Figure 4.11 but for a TWA with a buried stripe structure

region (i.e. the y-direction), and hence it is more sensitive to increasing d. Consequently, the polarisation sensitivities of the amplifier reduces with increasing d, which can also be seen from Figure 4.11. The results for a stripe geometry structure are shown in Figure 4.13. The refractive index and loss distribution in this structure arise principally because of lateral carrier diffusion into the surroundings [48] (Figure 4.14). Again, a reduction of polarisation sensitivity with increasing active layer thickness d is observed. This can be attributed to the improvement in the confinement of the quasi-TM mode with a

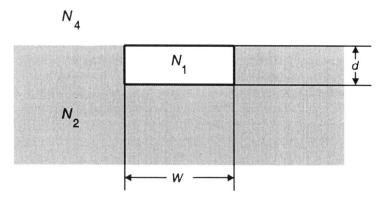

Figure 4.16   Refractive index distribution for a TWA with a buried channel (BC)

Table 4.4   Parameters used in analysing the polarisation sensitivities of different structures

---

(a) Buried heterostructure ($\lambda = 1.55\,\mu$m)

Active layer width:	$W = 3\,\mu$m
Active layer thickness:	$d = 0\text{-}1\,\mu$m
Length of amplifier:	$L = 300\,\mu$m
Refractive indices:	$N_1 = 3.524$
	$N_2 = N_3 = N_4 = N_5 = 3.169$
Net material gain coefficient:	$(g - \alpha_a) = 150\,\text{cm}^{-1}$
Loss coefficients:	$\alpha_2 = \alpha_3 = \alpha_4 = \alpha_5 = 50\,\text{cm}^{-1}$

(b) Stripe geometry ($\lambda = 1.55\,\mu$m)

Active layer width:	$W = 3\,\mu$m
Active layer thickness:	$d = 0\text{-}1\,\mu$m
Length of amplifier:	$L = 300\,\mu$m
Refractive indices:	$N_1 = 3.524$
	$N_2 = N_3 = 3.169$
	$N_4 = N_5 = 3.069$
Net material gain coefficient:	$(g - \alpha_a) = 150\,\text{cm}^{-1}$
Loss coefficients:	$\alpha_2 = 50\,\text{cm}^{-1}$
	$\alpha_3 = \alpha_4 = 100\,\text{cm}^{-1}$
	$\alpha_5 = 20\,\text{cm}^{-1}$

(c) Buried channel structure ($\lambda = 1.55\,\mu$m)

Active layer width:	$W = 3\,\mu$m
Active layer thickness:	$d = 0\text{-}1\,\mu$m
Length of amplifier:	$L = 300\,\mu$m
Refractive indices:	$N_1 = 3.524$
	$N_2 = N_3 = N_5 = 3.169$
	$N_4 = 3.269$
Wavelength:	$\lambda = 1.55\,\mu$m
Net material gain coefficient:	$(g - \alpha_a) = 150\,\text{cm}^{-1}$
Loss coefficients:	$\alpha_2 = \alpha_3 = \alpha_5 = 50\,\text{cm}^{-1}$
	$\alpha_4 = 100\,\text{cm}^{-1}$

---

thick active layer, similar to the improvement of the confinement of the TM mode as discussed by Saitoh and Mukai [9]. The results for a BC TWA in Figure 4.15 illustrate the impact of refractive index distribution on the polarisation characteristics of the TWA. The active region is formed by diffusing a channel into the substrate [30]. The striking feature of Figure 4.15 is the very small polarisation sensitivity of the structure as d approaches 1 μm. A virtually polarisation insensitive TWA with zero polarisation sensitivity is achieved for d larger than 0.55 μm. In addition, for small values of d, the signal gain for the quasi-TM mode is *larger* than that of the quasi-TE mode. This can be explained by the asymmetric refractive index step along x and y. Such asymmetry actually reduces the initial differences between the confinement of these two modes, and for a small active layer thickness the confinement for the quasi-TM mode is actually better than that for the quasi-TE mode. The subsequent improvement in confinement by increasing d further suppressed the gain dependence on the modal field distributions. Therefore it appears that a BC structure is useful in minimising the polarisation sensitivities in TWAs, although STM operation may not hold for this refractive index distribution [4, 16].

Effect of refractive index distribution

In order to investigate the effect of refractive index distribution on the polarisation characteristics of the hybrid modes, we introduce a parameter ΔN:

$$\Delta N = N_4 - N_2 \qquad (4.55)$$

We consider two circumstances, both with $N_2 = N_3$: (a) a symmetrical case with $N_4 = N_5$; and (b) an asymmetrical case with $N_4 \neq N_5$ but $N_2 = N_3 = N_5$. Notice the similarity of the refractive index distribution of the case (b) with that of a BC TWA (Figure 4.16). In the following analysis, $N_2 = N_3 = 3.169$, $d = 1 \mu$m and other parameters are as listed in Table 4.4(a), except for the values of N_4 and N_5, which vary according to ΔN.

Case (a): Symmetrical case.   As shown in Figure 4.17, the single pass gain G_s reduces as ΔN increases for both the quasi-TE and quasi-TM modes. Increasing the value of ΔN will lead to a reduction of confinement for both modes because of the increasing refractive index differences between the active region 1 and regions 4 and 5 (i.e. a larger refractive index step along the y-direction), which results in a smaller value of propagation constant β_0 in the perturbation analysis. The structure departs from the weakly guiding condition for bounded modes [1]. The reduction of confinement can be seen as roughly the same for negative values of ΔN, and as ΔN is increased the gain of the quasi-TE mode begins to fall faster than that of the quasi-TM mode. For $\Delta N = 0.18$ the gain is the same for both modes: this corresponds to a polarisation insensitive structure. Beyond that, the gain of the quasi-TM mode is larger than that of the quasi-TE mode. The more rapid fall in gain for the quasi-TE mode for positive values of ΔN can be explained by the more

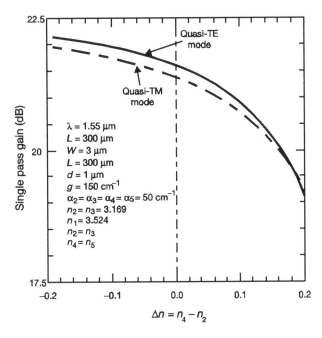

Figure 4.17 Effect of refractive index profile of the polarisation sensitivities of the amplifier gain for the fundamental quasi-TE and quasi-TM modes

significant reduction of confinement for the quasi-TE mode than that for the quasi-TM mode, which is due to the large refractive index step along the y-direction. As d is fixed at $1\,\mu m$, it can be seen that by using the combination of a thick active layer with an appropriate refractive index profile, the polarisation sensitivity for the TWA can be reduced below that obtained by increasing the active layer thickness only.

Case (b): Asymmetrical case. In this case only the refractive index in region 4 is varied. Therefore we would expect that the reduction of confinement for both modes would not be as rapid as with the previous case. This is confirmed in Figure 4.18, where the variation of G_s with ΔN is again calculated for this structure. Besides a slower rate of reduction in G_s with ΔN, observations similar to those made in the previous case can be made. Again, it can be seen that the confinement of the quasi-TE mode reduces more rapidly than that of the quasi-TM mode for positive values of ΔN, and again a cross-over between the two curves is observed. This cross-over point corresponds to the refractive index distribution for a polarisation insensitive TWA structure with given active region dimensions. The observations in both cases prove that the refractive index profile can play a significant role in controlling the polarisation sensitivities of TWAs. This is exactly the reason why separate confinement structures were suggested by Saitoh and Mukai [9] to further minimise the polarisation sensitivities of the TE and TM modes. However, our results do not indicate whether scheme (a) or (b) is better. In addition, the refractive index

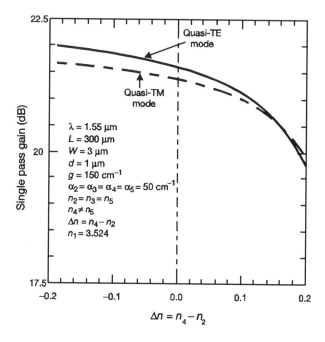

Figure 4.18   The same as Figure 4.17 but with an asymmetrical refractive index distribution

profile control has to be matched with the material characteristics and the fabrication processes of the amplifier [5]. An optimum way of controlling the refractive index profile has yet to be found.

Effect of active layer width

The transverse modal field distributions are affected by the active layer width W, as discussed in section 4.2 (see Table 4.2). Therefore the possibility exists of minimising the polarisation sensitivity for hybrid modes in a TWA by altering the active layer width W. Figure 4.19 shows some of the results of our investigation into this possibility. We have calculated the variation of the single pass gain G_s with the active layer width W for fundamental quasi-TE and quasi-TM modes of a TWA with different values of the active layer thickness d. The parameters used in the analyses are identical to those listed in Table 4.4(a), except that now the active layer width W is varied between $0\,\mu$m and $5\,\mu$m. Three values of the active layer thickness d are used: $0.5\,\mu$m, $1.0\,\mu$m and $1.5\,\mu$m, respectively. It can be seen that for all values of d, as W is reduced from $5\,\mu$m, the values of G_s for both hybrid modes stay nearly constant until a certain critical value of W is reached, where the values of G_s fall off rapidly for both hybrid modes. In this region the dimension of the structure is approaching the cut-off for the quasi-TE and quasi-TM modes for the wavelength of

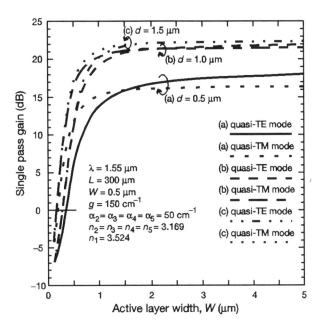

Figure 4.19 Variation of gain for the fundamental quasi-TE and quasi-TM modes with active layer width for a buried heterostructure TWA

interest. The effect of varying d can be seen as altering this critical value of W only. This can be explained by the fact that for a particular wavelength, the waveguide dimensions for which propagation is allowed is determined by the aspect ratio W/d [4] (see Figure 4.7). Hence as d changes, the value of W at which cut-off occurs will depend upon the aspect ratio.

We can see from Figure 4.19 that G_s stays fairly constant for both hybrid modes for the range of values of active layer widths that are well above cut-off. In addition, the differences between the gain for these two modes remain constant until the structure is operating in the cut-off region. It can be seen that altering W does not have a significant effect on the polarisation sensitivities of the *hybrid modes* of a TWA. The confinement of fields along the x-direction is less sensitive to waveguide dimensions compared with that along the y-direction. Changing the active layer thickness d will produce a more significant effect in minimising the polarisation sensitivities of the signal gain. Note finally from Figure 4.19 that the range of values of W for which the differences in G_s for both quasi-TE and quasi-TM modes are less than 1 dB increases with d. Hence a thick active layer can reduce the sensitivity of the gain differences between these two hybrid modes due to uncertainties in the active layer width. A final comment on the analysis of the polarisation sensitivities of SLAs using quasi-TE and quasi-TM modes: it can be seen from Figure 4.19 that the polarisation sensitivities of these higher-order modes are important *only* when W is large or, as revealed in Figure 4.11, for a fairly large value of d, when these modes will be excited. One can argue that it is possible to tailor the active layer width and thickness so that they satisfy the STM condition [28]. In practice, usually a small value of W is used such that STM can be retained, and increases d to reach an optimum polarisation sensitivity without disturbing the STM condition. This has an additional advantage of generating a more circular far-field pattern and hence enhancing the coupling efficiency with single-mode fibres. However, there are also advantages from increasing both the active layer width and the thickness. These include a lower electrical resistance and hence thermal dissipation [50], and also increases in the saturation output power [51] (see Chapter 5). In these circumstances, analysis of the polarisation characteristics of these hybrid modes becomes important.

4.4   IMPORTANCE OF TRANSVERSE MODAL FIELD PROPERTIES IN SLAs

We have seen in this chapter how to analyse the electric field distribution in a semiconductor laser amplifier by considering the rectangular dielectric waveguide formed in the active region. We have seen that for index guided structures the presence of gain and losses simply perturb the propagation constant of an identical passive waveguide by adding an imaginary part, which accounts for the amplification of the field along the amplifier. We have also seen that these fields are, in general, x–y dependent. Hence the signal gain becomes dependent on the modal field distribution in the amplifier. Failla *et al.* [27] discovered that this can affect the design of anti-reflection coatings as well, as confirmed by Vassallo [52, 53]. An experimental study on the polarisation sensitivities of SLAs will be presented in Chapter 7, together with the reported measurements on other SLA characteristics. Because the transverse modal field distribution ultimately determines the photon distribution within the amplifier, the modal fields should affect other amplifier characteristics like noise, output saturation power, etc. That is the pitfall in Saitoh and Mukai's analysis [9]. They have considered the polarisation sensitivity on gain, based only on the differences in optical confinement factors between the two polarisations, and utilise equation (4.38) in their subsequent analyses, probably for the sake of convenience. In practice, because the field distribution is different for different modes, not only will the optical confinement factor be affected, but also the photon distribution, and hence the recombination rate, the gain profile along the amplifier, etc. will be affected, and will ultimately affect the overall gain of the amplifier [54, 55]. Under such circumstances, calculating the gain dependence on polarisation with equation (4.38) based on the optical confinement factor only will not give a sufficiently accurate model. We will proceed, using the knowledge and tools acquired in this chapter, to form a systematic framework for analysing the different characteristics of laser amplifiers, taking account of the above physical processes. In Chapter 5 we will first examine how such a framework can be used to analyse the gain and saturation characteristics of SLAs.

4.5 REFERENCES

[1] A. W. Snyder and J. D. Love, *Optical Waveguide Theory*, Chapman & Hall, London, 1983.

[2] C. A. Balanis, *Advanced Engineering Electromagnetics*, John Wiley and Sons, New York, 1989.

[3] D. Marcuse, *Theory of Dielectric Optical Waveguides*, Academic Press, New York, 1974.

[4] M. J. Adams, *An Introduction to Optical Waveguides*, John Wiley and Sons, New York, 1981.

[5] G. H. B. Thompson, *Physics of Semiconductor Laser Devices*, John Wiley and Sons, New York, 1980.

[6] J. Buus, "The effective index method and its application to semiconductor lasers", *IEEE J. Quantum Electron.*, **QE-18**, No. 7, pp. 1083–1089, 1982.

[7] H. Ghafouri-Shiraz and C. Y. J. Chu, "Analysis of waveguiding properties of travelling-wave semiconductor laser amplifiers using perturbation technique", *Fiber and Integrated Opt.*, **11**, pp. 51–70, 1992.

[8] M. J. O'Mahony, "Semiconductor laser optical amplifiers for use in future fiber systems", *IEEE J. Lightwave Technol.*, **LT-6**, No. 4, pp. 531–544, 1988.

[9] T. Saitoh and T. Mukai, "Structural design for polarization-insensitive travelling-wave semiconductor laser amplifiers", *Opt. Quantum Electron.*, **21**, pp. S47–S48, 1989.

[10] H. Ghafouri-Shiraz and C. Y. J. Chu, "Analysis of polarization sensitivity in semiconductor laser amplifiers using the perturbation method", *Trans. IEICE Jpn, National Convention Record*, Paper No. C-188, 1991.

[11] R. E. Collin, *Field Theory of Guided Waves*, McGraw-Hill, New York, 1960.

[12] A. Yariv, *Optical Electronics*, 3rd edition, Holt–Saunders, 1985.

[13] A. S. Sudbø, "Why are accurate computations of mode fields in rectangular dielectric waveguides difficult?", *J. Lightwave Technol.*, **LT-10**, No. 4, pp. 418–419, 1992.

[14] P. M. Morse and H. Feshbach, *Methods of Theoretical Physics: Part I*, McGraw-Hill, New York, 1953.

[15] N. H. G. Baken, M. B. J. van Diemeer, J. M. Splunter and H. Blok, "Computational modelling of diffused channel waveguides using a domain integral equation", *IEEE J. Lightwave Technol.*, **LT-8**, No. 4, pp. 576–586, 1990.

[16] W. P. Huang, "Polarization corrections of dispersion characteristics of optical waveguides", *Opt. Lett.*, **15**, No. 19, pp. 1052–1054, 1990.

[17] M. Born and E. Wolf, *Principles of Optics*, 6th edition, Pergamon Press, Oxford, UK, 1983.

[18] S. G. Mikhlin and K. L. Smolitskiy, *Approximate Methods for Solution of Differential and Integral Equations*, American Elsevier Publishing Co., 1967.

[19] A. Yariv and P. Yeh, *Optical Waves in Crystals*, John Wiley and Sons, New York, 1984.

[20] T. Ikegami, "Reflectivity of a mode at facet and oscillation made in double-heterostructure injection laser", *IEEE J. Quantum Electron.*, **QE-8**, No. 6, pp. 470–476, 1972.

[21] T. E. Rozzi, and G. H. in'tVeld, "Variational treatment of the diffraction at the facet of d.h. lasers and of dielectric millimeter wave antennas", *IEEE Trans. Microwave Theory Technol.*, **MTT-28**, No. 2, pp. 61–73, 1980.

[22] L. Horn and C. A. Lee, "Choice of boundary conditions for rectangular dielectric waveguides using approximate eigenfunctions separable in x and y", *Opt. Lett.*, **15**, No. 7, pp. 349–350, 1990.

[23] P. C. Kendall, M. J. Adams, S. Ritchie and M. J. Robertson, "Theory for calculating approximate values for the propagation constants of an optical rib waveguide by weighting the refractive indices", *IEE Proc.*, **134**, Part A, No. 8, pp. 699–702, 1987.

[24] M. J. Robertson, P. C. Kendall, S. Ritchie, P. W. A. McIlroy and M. J. Adams, "The weighted index method: a new technique for analysing planar optical waveguides", *IEEE J. Lightwave Technol*, LT-7, No. 12, pp. 2105–2111, 1989.

[25] V. J. Menon, S. Bhattacharjee and K. K. Dey, "The rectangular dielectric waveguide revisited", *Opt. Commun.*, **85**, No. 5 and 6, pp. 393–396, 1991.

[26] E. A. J. Marcatilli, "Dielectric rectangular waveguides and directional couplers for integrated optics", *Bell Syst. Technol. J.*, **48**, pp. 2071–2102, 1969.

[27] A. G. Failla, G. P. Brava and I. Montrosset, "Structural design criteria for polarization insensitive semiconductor optical amplifiers", *J. Lightwave Technol.*, **LT-8**, No. 3, pp. 302–308, 1990.

[28] H. Ghafouri-Shiraz, "Single transverse mode condition in long wavelength SCH semiconductor laser diodes", *Trans. IEICE*, **E70**, No. 2, pp. 130–134, 1987.

[29] G. P. Agrawal and N. K. Dutta, *Long-wavelength Semiconductor Lasers*, Van-Nostrand Reinhold, New York, 1986.

[30] W. Huang, H. A. Haus and H. N. Yoon, "Analysis of buried-channel waveguides and couplers: scalar solution and polarization correction", *IEEE J. Lightwave Technol.*, **LT-8**, No. 5, pp. 642–648, 1990.

[31] Y. H. Cheung and W. G. Lin, "Investigation of rectangular dielectric waveguides: an iteratively equivalent index method", *IEE Proc.*, **137**, Part J, No. 5, pp. 323–329, 1990.

[32] K. S. Chiang, "Analysis of rectangular dielectric waveguides: effective index with built-in perturbation correction", *Electron. Lett.*, **28**, No. 4, pp. 388–390, 1992.

[33] C. H. Henry and B. H. Verbeck, "Solution of the scalar wave equation for arbitrarily shaped dielectric waveguides by two-dimensional Fourier analysis", *IEEE J. Lightwave Technol.*, **LT-7**, No. 32 pp. 308–313, 1989.

[34] C. H. Henry and Y. Shani, "Analysis of mode propagation in optical waveguide devices by Fourier expansion", *IEEE J. Quantum Electron.*, **QE-27**, No. 3, pp. 523–530, 1991.

[35] G. B. Hocker and W. K. Burns, "Mode dispersion in diffused channel waveguides by the effective index method", *Appl. Opt.*, **16**, No. 1, pp. 113–118, 1977.

[36] H. A. Haus and S. Kawakami, "On the excess spontaneous emission factor in gain-guided laser amplifiers", *IEEE J. Quantum Electron.*, **QE-21**, No. 1, pp. 63–69, 1985.

[37] H. A. Haus, *Wave and Fields in Optoelectronics*, Prentice-Hall, New York, 1984.

[38] J. E. Goell, "A circular-harmonic computer analysis of rectangular dielectric waveguides", *Bell Syst. Tech. J.*, **48**, No. 7, pp. 2133–2160, 1969.

[39] J. Buus, "Principles of semiconductor laser modelling", *IEE Proc.*, **132**, Part J, No. 1, pp. 42–51, 1985.

[40] S. Banerjee and A. Sharma, "Propagation characteristics of optical waveguiding structures by direct solution of the Helmholtz equation for total fields", *J. Opt. Soc. Am. A*, **6**, No. 12, pp. 1884–1894, 1989.

[41] T. Young, "Finite element modelling of a polarization independent optical amplifier", *IEEE J. Lightwave Technol.*, **LT-10**, No. 5, pp. 626–633, 1992.

[42] C. N. Capsalis, J. G. Fikjoris and N. K. Uzunoglu, "Scattering from an abruptly terminated dielectric slab waveguide", *IEEE J. Lightwave Technol.*, **LT-3**, No. 2, pp. 408–415, 1985.

[43] W. K. Uzungalu and C. N. Capsalis, "Diffraction from an abruptly terminated dielectric slab waveguide in the presence of a cylindrical scatterer", *IEEE J. Lightwave Technol.*, **LT-4**, No. 4, pp. 405–414, 1986.

[44] E. W. Kolk, N. H. G. Baken and H. Blok, "Domain integral equation analysis of integrated optical channel and ridge waveguide in stratified media", *IEEE Trans. Microwave Theory Technol.*, **38**, No. 1, pp. 78–85, 1990.

[45] H. J. W. M. Hoekstra, "An economic method for the solution of the scalar wave equation for arbitrarily shaped optical waveguides", *IEEE J. Lightwave Technol.*, **LT-8**, No. 5, pp. 789–793, 1990.

[46] G. W. Wen, "Steepest-descent approximation theory for guided modes of weakly guiding waveguides and fibres", *J. Opt. Soc. Am. A*, **8**, No. 2, pp. 295–302, 1991.

[47] A. Yariv, *Quantum Electronics*, 3rd edition, John Wiley and Sons, New York, 1989.

[48] M. Cross and M. J. Adams, "Waveguiding properties of stripe-geometry double heterostructure injection lasers", *Solid State Electron.*, **15**, pp. 919–921, 1972.

[49] C. Y. J. Chu and H. Ghafouri-Shiraz, "Structural effects on polarization sensitivity of travelling-wave semiconductor laser amplifiers", *3rd Bangor Communication Symposium*, United Kingdom, pp. 19–22, 1991.

[50] J. C. Simon, "GaInAsP semiconductor laser amplifiers for single-mode fiber communications", *J. Lightwave Technol.*, **LT-5**, No. 9, pp. 1286–1295, 1987.

[51] Y. Mukai, Y. Yamamoto and T. Kimura, "Optical amplification by semiconductor lasers", *Semiconductors and Semimetals*, **22**, Part E, pp. 265–319, Academic Press, London, 1985.

[52] C. Vassallo, "Theory and practical calculation of antireflection coatings on semiconductor laser diode optical amplifiers", *IEE Proc.*, **137**, Part J, No. 4, pp. 193–202, 1990.

[53] C. Vassallo, "Some numerical results on polarization insensitive 2-layer antireflection coatings for semiconductor optical amplifiers", *IEE Proc.*, **137**, Part J, No. 4, pp. 203–204, 1990.

[54] D. Marcuse, "Computer model of an injection laser amplifier", *IEEE J. Quantum Electron.*, **QE-19**, No. 1, pp. 63–73, 1983.

[55] M. J. Adams, J. V. Collins and I. D. Henning, "Analysis of semiconductor laser optical amplifiers", *IEE Proc.*, **132**, Part J, pp. 58–63, 1985.

ANALYSIS AND MODELLING OF SEMICONDUCTOR LASER AMPLIFIERS: GAIN AND SATURATION CHARACTERISTICS

5.1 INTRODUCTION

In Chapter 2, a simple two-level system was used to explain the fundamental principles and properties of an optical amplifier. The operational principles of a semiconductor laser amplifier (SLA) is more complex than that anticipated by the two-level model. For a deeper insight into SLAs, a more thorough understanding of the detailed mechanisms in SLAs is essential. Some of the basic physics of SLAs were explored in Chapter 3, where different structures and applications of SLAs were also discussed. It was shown that to further evaluate the gain and saturation characteristics of SLAs, the following quantities have to be known:

(i) the transverse modal field distribution across the active region of the amplifier,

(ii) the material gain coefficient profile along the amplifier; and

(iii) other external factors such as pumping rate (i.e. magnitude of injection or bias current), input optical power, etc.

The first two factors are determined by both the material characteristics and the structural parameters of the SLA. We have examined how (i) can be accomplished in the previous chapter. In this chapter we will proceed to analyse (ii) and, by incorporating (iii) into the analysis, construct a model that can be used to analyse the various performance characteristics of SLAs.

In practice, the material gain coefficient in the active region cannot be determined very easily. This is because of the inter-coupling effect between the optical processes (determined by the photon rate equation) in the active p–n junction [1, 2]. When optical signals propagate across the amplifier, they interact with the active region via stimulated emissions and induced absorptions. The number of photons generated by stimulated emissions and lost via induced absorptions is affected by the carrier density in the conduction band (i.e. the electron density) and the valence band (i.e. the hole density); see subsection 2.2.5, equation (2.14). However, such optical processes will be accompanied by recombinations of carriers, thus reducing the population inversion level in the amplifier. As we shall see later, this can be described by two simultaneous rate equations, which have to be solved in order to determine the exact material gain coefficient profile along the amplifier.

This chapter is organised as follows. First we examine how to analyse the gain and saturation characteristics of a SLA with a uniform material gain coefficient profile (hereafter for simplicity we call it a *uniform gain profile*), since this is the simplest analysis and can be used in some cases for approximation purposes [3]. Two different approaches to analysing SLAs will then be reviewed, where their merits and shortcomings will be discussed. A third approach proposed by Chu and Ghafouri-Shiraz [4] is reported next, which is based on the transfer matrix method (TMM) [5]. Using this approach, a more robust and efficient technique can be used to analyse SLAs with both relatively simple (e.g. buried heterostructures) and also more complicated structures (e.g. two-section amplifiers [6]). This technique of analysis also allows us to construct an *equivalent circuit model* to model SLAs. Finally, we will apply this model to examine the gain and saturation characteristics of SLAs, with particular reference to structural effects on device characteristics and system consideration.

5.2 ANALYSIS OF SEMICONDUCTOR LASER AMPLIFIERS WITH A UNIFORM GAIN PROFILE

The simplest way to analyse the gain and saturation characteristics is to assume that the material gain coefficient is the same along the entire amplifier. In other words, we assume that the effect of increasing the field amplitude (due to optical amplification) along z will not affect the carrier density n, and hence the material gain coefficient g_m, significantly. The validity of this commonly used assumption will be discussed later in subsection 5.2.3. But first we will see how to analyse the gain and saturation characteristics of a SLA under such an assumption. This will provide us with qualitatively accurate information about the basic characteristics of SLAs.

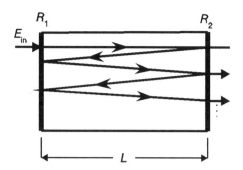

Figure 5.1 Model used to analyse the signal gain of a Fabry–Perot amplifier

5.2.1 Amplifier gain formulation in semiconductor laser amplifiers

Active Fabry–Perot formulation

We have seen how the material gain coefficient g_m can be used to evaluate the single pass gain G_s of a SLA using equation (4.37) of Chapter 4. To calculate the overall amplifier gain G, we must take into account the feedback, if any, provided by the finite end-facet reflectivity of the structure (see subsection 3.4.1 of Chapter 3). The structure can be represented by the simplified model of a Fabry–Perot (FP) cavity illustrated in Figure 5.1, which shows a SLA with length L and power reflection coefficients R_1 and R_2, respectively, at the end facets. The amplifier is assumed to have a uniform gain profile. The cavity resonance frequency f_0 is given by [7]

$$f_0 = \frac{mc}{2N_g L} \tag{5.1}$$

where m is an integer (i.e. the longitudinal mode number), c is the velocity of light in free space and N_g is the group effective refractive index of the active region (i.e. $N_g = v_g/c$, v_g is the group velocity) [8–10], which can be found via the modal field analysis described in the previous chapter.

Consider an optical signal with electric field E_{in} of amplitude E_0 and angular frequency $\omega = 2\pi f_0$ incident onto the SLA. The longitudinal propagation constant β_z of the field in the active region is given by (see subsection 4.2.3, equation (4.26))

$$\beta_z = \frac{2\pi N_g}{\lambda} \tag{5.2}$$

where λ is the signal wavelength. The output electric field E_{out} for a uniform gain profile along the amplifier can be found as the sum of multiple reflected signals transmitted from the amplifier cavity due to reflections of R_1 and R_2, which is given by [7]

$$E_{out} = \sqrt{(1 - R_1)(1 - R_2)G_s}\, E_0 \, \exp(-j\beta_z L)$$
$$\times\, [1 + \sqrt{R_1 R_2}\, G_s \, \exp(-2j\beta_z L) + R_1 R_2 G_s^2 \, \exp(-4j\beta_z L) + \ldots] \quad (5.3)$$

Using the sum for an infinite geometric progression, equation (5.3) can be simplified to

$$E_{out} = \frac{\sqrt{(1 - R_1)(1 - R_2)G_s}\, E_0 \, \exp(-j\beta_z L)}{1 - G_s\sqrt{R_1 R_2}\, \exp(-2j\beta_z L)} \quad (5.4)$$

The measured power of the amplifier output is related to the square of the magnitude of E_{out}, and from equation (4.18) it can be shown that

$$|E_{out}|^2 = \frac{(1 - R_1)(1 - R_2)G_s E_0^2}{(1 - G_s\sqrt{R_1 R_2})^2 + 4G_s\sqrt{R_1 R_2}\, \sin^2(\beta_z L)} \quad (5.5)$$

Therefore the amplifier gain is given by

$$G = \left|\frac{E_{out}}{E_{in}}\right|^2 = \frac{(1 - R_1)(1 - R_2)G_s}{(1 - G_s\sqrt{R_1 R_2})^2 + 4G_s\sqrt{R_1 R_2}\, \sin^2(\beta_z L)} \quad (5.6)$$

It can be seen from the above equation that the amplifier gain G is a periodic function of β_z (i.e. with wavelength λ and frequency f as well). The maximum signal gain occurs when the sine term goes to zero, or

$$f = \frac{pc}{2N_g L} = \frac{pf_0}{m} = qf_0 \quad (5.7)$$

where p and q are integers. Equation (5.7) implies that G is maximum when the input signal frequency is an integral multiple of the cavity resonance frequency. The corresponding value of the gain G_{max} is given by

$$G_{max} = \frac{(1 - R_1)(1 - R_2)G_s}{(1 - G_s\sqrt{R_1 R_2})^2} \quad (5.8)$$

The amplifier gain is minimum when the sine term is unity in equation (5.16). In this case the frequency f is given by

$$f = \frac{(2p + 1)c}{4N_g L} \quad (5.9)$$

This corresponds to a phase mismatch of π radians between the signal frequency and the cavity resonances. The minimum value of G is given by

$$G_{min} = \frac{(1 - R_1)(1 - R_2)G_s}{(1 + G_s\sqrt{R_1 R_2})^2} \quad (5.10)$$

The level of the ripples in the amplifier gain spectrum is described by the ratio between G_{max} and G_{min} The parameter ξ can be used to indicate this ratio [11, 12]. Using equations (5.8) and (5.10) we have

$$\xi = \left(\frac{1 + G_s\sqrt{R_1 R_2}}{1 - G_s\sqrt{R_1 R_2}}\right)^2 \qquad (5.11)$$

Equation (5.11) provides a method of measuring the single pass gain if the facet reflectivities are known [11]. The ripple level at the amplifier output is measured by scanning the amplified spontaneous emission spectrum, and then G_s can be calculated using the above equation. A knowledge of G_s allows us to calculate G using equation (5.6). To facilitate this in the analysis, Mukai and Yamamoto [12] suggested an alternative form of equation (5.6) by noting the periodicity of the gain function, and the fact that the term $\beta_z L$ actually corresponds to the phase mismatch between the signal and cavity resonances [14]

$$G = \frac{(1 - R_1)(1 - R_2)G_s}{(1 - G_s\sqrt{R_1 R_2})^2 + 4G_s\sqrt{R_1 R_2}\ \sin^2\left(\dfrac{2\pi(f - f_0)N_g L}{c}\right)} \qquad (5.12)$$

where f_0 is the resonance frequency of the cavity. Notice the similarity between equation (5.12) and equation (5.6) except for the replacement of the β_z term by a frequency term. It should be noted that from both equations (5.12) and (5.6), $G = G_s$ when $R_1 = R_2 = 0$ (i.e. when the SLA is a travelling-wave amplifier (TWA)). Measuring ξ also offers a method for measuring the facet reflectivities [15, 16]. Because G_s is related to the injection current i, we can change i to obtain a variation of ξ, and from that R_1 and R_2 can be calculated easily from equation (5.11) if $R_1 = R_2$.

Equation (5.11) shows that the ripple level ξ can be reduced by reducing the facet reflectivities R_1 and R_2. This is illustrated in Figure 5.2, where we have analysed a Fabry–Perot amplifier (FPA) with different facet reflectivities and single pass gain G_s. A FPA with $L = 300\,\mu m$, $N_g = 3.524$ and $\lambda = 1.55\,\mu m$ is analysed for (i) $R_1 = R_2 = 0$, (ii) $R_1 = R_2 = 0.3$ and (iii) $R_1 = R_2 = 0.5$, respectively. The resulting signal gain is plotted against $(f - f_0)$, calculated using equation (5.12). In these calculations we have neglected the dependence of the material gain on the signal wavelength. The effective signal bandwidth is also increased, and a flat and uniform spectrum is obtained when $R_1 = R_2 = 0$, which corresponds to a TWA. The gain can be maximised when both R_1 and R_2 are non-zero and the FPA is tuned to resonance. In practice, R_1 and R_2 are often finite (see the discussion in subsection 3.4.1 in Chapter 3) and hence the amplifier gain spectrum is seldom flat. The gain of the amplifier can then be maximised by tuning the cavity resonance of the amplifier to match the input signal frequency. This is usually achieved by changing the temperature of the amplifier [17].

In Figure 5.3 we have plotted the variation of ξ with G_s for $R_1 = R_2 = 0.3$. The magnitude of ξ increases rapidly as G_s increases, and hence a FPA is more sensitive to a frequency mismatch if G_s is high, making it unsuitable as a stable high-gain power amplifier [13]. Indeed, if G_s is increased such that the denominator of equation (5.11) becomes zero, the FPA reaches the lasing threshold and oscillations occur. This is the familiar threshold condition in semiconductor lasers [1, 7, 18]. The above derivations of the signal gain of

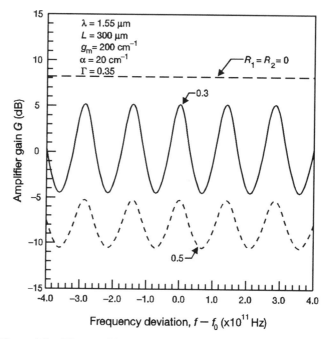

Figure 5.2   The amplifier gain spectrum of a Fabry–Perot amplifier

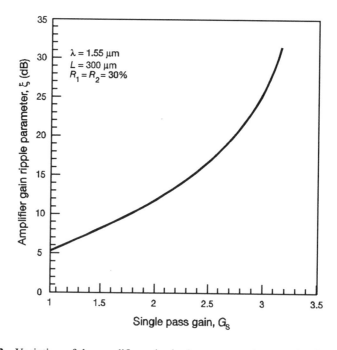

Figure 5.3   Variation of the amplifier gain ripple parameter ξ versus the single-pass gain G_s of a Fabry–Perot amplifier

SLAs, based on the Fabry–Perot approach, have taken into account the phase mismatch due to cavity resonances. Hence, it is also known as an active FP formulation [19]. Another possible method is to apply photon statistics [12, 20]. In what follows, we will describe this latter approach briefly.

Photon statistics formulation

According to photon statistics, the signal gain G_{co} is defined as the ratio of the increment of the number of photons extracted from the output mirror to the number of photons injected into the input mirror [19, 21]. If the signals are tuned to the mth cavity resonance to achieve maximum amplifier gain, G_{co} can be defined as

$$G_{co} = \frac{N_{p,m}((P_m)_{in} \neq 0) - N_{p,m}((P_m)_{in} = 0)}{\tau_p (P_m)_{in}} \tag{5.13}$$

where τ_p is the total photon lifetime due to scattering loss, $(P_m)_{in}$ is the number of injected photons to the mth cavity resonance of the FP cavity per second, and $N_{p,m}$ is the total number of photons in the mth cavity resonance given by [12]

$$N_{p,m} = \frac{\Gamma E_{cf,m} + (P_m)_{in}}{\tau_p^{-1} - \Gamma G_m} \tag{5.14}$$

where Γ is the optical confinement factor, G_m is the *stimulated emission coefficient*, which is equal to the stimulated emission rate, and $E_{cf,m}$ is the *spontaneous emission probability* [11, 12]. These two quantities are related to the band structure of the material. The subscript m indicates that these parameters are calculated with respect to the mth cavity resonance frequency. The numerator in equation (5.13) involves two terms. The first term is the total number of photons in the mth mode, including signal, stimulated emissions and amplified spontaneous emissions. The second term with $(P_m)_{in} = 0$ isolates out the signal photons, and includes those photons due to amplified spontaneous and stimulated emissions induced by amplified spontaneous emissions only. Hence, the numerator of equation (5.13) represents the total number of photons due to the injected signal and those due to stimulated emissions induced by the signal. The form of equation (5.13) excludes the effect of finite coupling losses at the input and output mirror facets with power reflection coefficients of R_1 and R_2, respectively (see Figure 5.1). To take account of these, we introduce two efficiency parameters [12]. The injection efficiency η_{in} which is given by

$$\eta_{in} = \frac{(1 - R_1) G_s}{1 + R_1 G_s} \tag{5.15}$$

and the extraction efficiency η_{out}, given by

$$\eta_{\text{out}} = \frac{\tau_p}{\left(\dfrac{c}{LN_g}\right) \cdot \ln\left(\sqrt{\dfrac{1}{R_2}}\right)} \tag{5.16}$$

The derived signal gain G using this approach is given by [12]

$$G = \eta_{\text{in}} G_{\text{c0}} \eta_{\text{out}} = \frac{N_{\text{p},m}((P_m)_{\text{in}} \neq 0) - N_{\text{p},m}((P_m)_{\text{in}} = 0)}{\left(\dfrac{c}{LN_g}\right) \cdot \ln\left(\sqrt{\dfrac{1}{R_2}}\right) \cdot (P_m)_{\text{in}}} \cdot \frac{(1 - R_1)G_s}{1 + R_1 G_s} \tag{5.17}$$

Comparisons between the two formulations

We have described two different mathematical descriptions of G: equations (5.6) and (5.12) using the active FP approach, and equation (5.17) using the photon statistics approach. An inspection of the different forms of equations (5.17) with (5.12) reveals that the photon statistics formulation does not take into account the frequency dependence of the amplifier gain spectrum (see Figure 5.1), and requires additional information about the amplified spontaneous emission in order to calculate $N_{\text{p},m}$. Mukai and Yamamoto [12, 19] found in their investigations that these two approaches gave different results. Marcuse [3] argued that the photon statistics approach assumes that one can analyse the characteristics of a SLA accurately by ignoring the spatial variation of the number of photons along the cavity, and hence the use of the total number of photons to calculate G can be justified. He pointed out that this requires a uniform electric field distribution along the cavity. Whereas this assumption is approximately true in a FP cavity, the field distribution becomes increasingly non-uniform as the facet reflectivities of the SLA reduce. Hence, the validity of equation (5.17) becomes questionable in analysing SLAs with low facet reflectivities (e.g. in TWAs). Cassidy [22] compared these two approaches critically and found that the active FP approach is more appropriate, because the phase information of the fields inside the cavity can be retained. In order to improve the accuracy in the analysis of the characteristics of SLAs, the active FP approach should be adopted.

5.2.2 Gain saturation formulation in semiconductor laser amplifiers

In the previous section we examined how to analyse the amplifier gain for SLAs. However, a careful examination of the previous mathematical expressions reveals that the material gain (and hence the amplifier gain) is not affected by the input signal power for a fixed injection current, which is contrary to experimental observations [19], as well as the qualitative prediction using a simple two-level system as explained in subsection 2.3.3 in Chapter 2. In SLAs, as the signal photons propagate across the amplifier, stimulated emissions can also change the population inversion level. The change in population level is dynamic, and the competition between stimulated emissions and recombinations can be described by the rate equations similar to equations (2.12) and (2.15) [12, 19]. Usually these rate equations have to be solved by numerical iterations [18]. The saturation output power for a SLA with a particular bias or injection current can then be found by plotting the amplifier gain calculated from these rate equations against optical power, and noting the $-3\,$dB point from the unsaturated gain value on the plot [23]. However, an analytical form of gain saturation in terms of the output optical intensity can be obtained under the uniform gain profile assumption [13, 24, 25].

The optical intensity $I(z)$ is allowed to vary, as in the active FP formulation. This gives the familiar expression of material gain $g_m(z)$ at position z. That is

$$g_m(z) = \frac{g_0}{1 + \dfrac{I(z)}{I_{Sat}}} \tag{5.18}$$

where g_0 is the unsaturated material gain coefficient and I_{Sat} is the saturation intensity [13, 26]. Details of the derivation of equation (5.18) can be found in Section 5.7. If g_m is measured at the output of the amplifier, i.e. $z = L$, then I_{Sat} refers to the output saturation intensity. The exact value of I_{Sat} can be found by solving the rate equations [2]. Yamamoto *et al.* [26] have found that I_{Sat} is given by

$$I_{Sat} = \frac{hf}{A\tau_{sp}} = \frac{chf}{\beta N_g V_0} \tag{5.19}$$

where A is the differential gain coefficient described in equation (3.5) of sub-section 3.2.2, τ_{sp} is the carrier recombination lifetime due to spontaneous emission only, N_g is the group effective refractive index of the amplifying medium, $V_0 = V/\Gamma$ is the optical mode volume, V is the active region actual volume, Γ is the confinement factor and β is the spontaneous emission coefficient, which is defined as the fraction of spontaneous emission power coupled to the signal [27–30]. Notice that I_{Sat} depends mainly on the material characteristics. Thus the saturation output power, given by $P_{Sat} = I_{Sat} \cdot W \cdot d$ (where W and d are, respectively, the width and thickness of the active region) can be increased by increasing W and/or d [2, 13]. A typical value of β is around 10^{-4} for $1.3\,\mu$m and $1.55\,\mu$m wavelength SLAs [23]. As we have been dealing with photons in solving the rate equations (see Section 5.7), we can approximate $I(L)$ by

$$I(L) = \frac{chf}{N_g V_0} \langle N_p \rangle \tag{5.20}$$

where $\langle N_p \rangle$ is the average number of photons at the output of the amplifier. Using equations (5.19) and (5.20), we can re-express equation (5.18) by [13]

$$g_m(L) = \frac{g_0}{1 + \beta \langle N_p \rangle} \tag{5.21}$$

at the amplifier output $z = L$. Therefore the gain saturation is related to the amount of spontaneous emissions coupled to the signal as well as the signal power (i.e. $\langle N_p \rangle$) itself. In fact, the material gain reduces to $g_0/2$ when the number of photons $\langle N_p \rangle = 1/\beta$, as revealed by equation (5.21). It should be pointed out that the above expressions are not only valid under the assumption of a uniform gain profile along the amplifier, but also, as shown in Section 5.7 and also in Chapter 6, they can be applied at any point z in the amplifier. The SLA starts to saturate when the material gain coefficient at a particular distance z begins to reduce significantly from its unsaturated value.

Another interesting aspect of gain saturation in SLAs is its homogeneous nature [31]. We have illustrated in Chapter 3 (Figure 3.4) that the effect of increasing g_{peak} on the material gain spectrum is to shift the whole spectrum homogeneously upwards. The same homogeneous behaviour can be observed when the amplifier saturates. When saturation occurs for a particular wavelength in the material gain spectrum due to high incident power, the material gain coefficients for other wavelengths across the entire spectrum also saturate simultaneously. This occurs because saturation is due to the total photon density (i.e. total power of signals and spontaneous emissions inside the cavity according to equation (5.21)), which explains the above phenomenon: saturation occurs homogeneously because it depends on the total amount of photons for all wavelengths that exist in the cavity.

5.2.3 Appraisal on using a uniform gain profile in analysing SLAs

We have compared the underlying assumptions in the different approaches in analysing the gain and saturation characteristics, which are described above, in Table 5.1. It can be seen that the photon statistics approach is physically unrealistic, where both the gain profile and the signal fields (hence photon density) are assumed to be uniform. In the active FP formulation, we have assumed that the material gain g_m is independent of z, although the variation of the field has been accounted for by considering the phase of the fields at the facets [22]. It has been shown that in semiconductor lasers with Fabry–Perot cavities, the electric field distribution is fairly uniform in the cavity along z because of the relatively high Q-factor (due to the relatively high value of facet reflectivities) of the optical cavity (which creates a standing-wave pattern) [10], and hence g_m can be assumed to be z-independent without significant loss in

Table 5.1 Comparisons between different approaches of analysing SLAs

Analysis method	Is gain coefficient independent of z?	Is field amplitude (photon density) independent of z?
Photon statistics/rate equation	Yes	Yes
Active FP formulation	Yes	No
Travelling-wave equations	No	No

accuracy [1, 32], but note the critical comment concerning the loss in phase information made by Cassidy [22]. On the other hand, for other laser structures (i.e. distributed feedback lasers [33]) and most laser amplifiers [3, 23], where the facet reflectivities are fairly low, this assumption will not hold. A more accurate analysis of SLAs will require consideration of the interaction between the injected carrier density, the material gain and the electric field amplitude point by point within the cavity. Such a modelling of SLAs will be discussed in the following sections.

5.3 GENERAL ANALYSIS OF SEMICONDUCTOR LASER AMPLIFIERS (A BRIEF REVIEW)

In the previous section the amplifier gain and the saturation characteristics (i.e. the dependence of the amplifier gain on the optical intensity) for a SLA were analysed by assuming a uniform gain profile. But, to calculate these quantities, the value of the material gain coefficient first has to be found. This can be obtained conveniently by solving the modified multi-mode rate equations based on photon statistics [1, 12], the procedures for which will be described in subsection 5.3.1. However, as highlighted in subsection 5.2.3, using photon statistics in analysing the characteristics of SLAs assumes implicitly that both the gain and the electric field profile are independent of z [3]. This is often not true in SLAs, and hence alternative methods for analysing the gain profile and subsequently the amplifier gain and saturation characteristics have to be derived. A commonly used method, known as the travelling-wave equation method, will be described in subsection 5.3.2 [3, 23].

5.3.1 Analysis using rate equations

This approach is a modification of the well-known laser multi-mode rate equations [1, 34] by including an input signal term in the photon density rate equation [20]. In this case, the time derivatives of the photon density S_m and carrier density n can be expressed by the following two equations;

$$\frac{dn}{dt} = P_c - R - \sum_m \Gamma G_m S_m \tag{5.22}$$

$$\frac{dS_m}{dt} = -\frac{S_m}{\tau_p} + \Gamma G_m S_m + \Gamma E_{cv,m} + S_{in,m} \tag{5.23}$$

In the above equations P_c is the pumping rate due to the injection current, R is the recombination rate of the carriers and $S_{in,m}$ is the photon density injected by the incident optical signal per second. Other symbols have already been defined in subsection 5.2.1. Notice that the rate equation for the carriers, equation (5.22), describes fluctuations in the carrier density n due to all of the optical modes in the SLA, whereas that for the photon density, equation (5.23), describes the fluctuations in the photon density for each optical mode.

Mathematical models for R can be found in subsection 3.2.2 of Chapter 3. For SLAs operating with long wavelengths of $1.3\,\mu m$ and $1.55\,\mu m$ the superlinear model (equation (3.10)) has to be used because of the significance of Auger recombinations [15]. Then the carrier density can be solved for the above equations, either for transient analysis [35] or for steady-state operation, where the left-hand side of equation (5.22) will be set to zero for all values of m [12]. In both analyses the solutions are found by a self-consistency analysis, similar to that for lasers [18, 36]. Once the carrier density has been found, the gain coefficient g_m can be derived from equations (3.5) or (3.6) in Chapter 3, and hence the signal gain can be calculated by either the photon statistics formulation (subsection 5.2.1) by assuming both a uniform carrier and photon distribution along the amplifier, or by using the active FP formulation described in subsection 5.2.1. The gain saturation characteristics can subsequently be found following the procedures described in subsection 5.2.2.

5.3.2 Analysis using travelling-wave equations

The above analysis using rate equations assumes that both the photon density and the material gain (and hence the carrier density) are independent of z (see Table 5.1). In order to take into account the spatial dependence of the carrier and photon densities, Marcuse [3] modified the rate equations for the mth mode as

$$\frac{dn(z,\,t)}{dt} = \frac{j}{ed} - R - \sum_m \Gamma G_m(z) S_m(z) \tag{5.24}$$

$$\frac{dS_m(z,\,t)}{dt} = \frac{\beta n(z)}{\tau_{sp}} - \frac{S_m(z)}{\tau_p} + \Gamma G_m(z) S_m(z) \tag{5.25}$$

where S_m is the photon density of the mth guided mode, including both photons injected by the input signal, and photons generated by stimulated emissions and spontaneous emission, τ_{sp} is the radiative carrier recombination lifetime due to spontaneous emissions, j is the injection current density, d is the thickness of the active layer, e is the electronic charge ($=1.6 \times 10^{-19} C$) and β is the fraction of spontaneous emissions coupled to the mth mode. In deriving the above equations the current injection efficiency is assumed to be 100%. For an energy bandwidth of $\Delta E = h\Delta f$ (i.e. an optical frequency bandwidth of Δf), we have

$$\sum_m G_m S_m = \frac{c}{N_g \Delta E} \int_0^{\Delta E} g_m(z,\,E) S(z,\,E)\, dE \tag{5.26}$$

where $S(z,\,E)$ is the total photon density at z with photon energy E and $g_m(z,\,E)$ is the gain coefficient of the material at z with energy E taking into account the spectral dependence. Notice that $S(z,\,E)$ includes both spontaneous emissions and signal photons. By assuming a steady-state operation, equation (5.24) can be rewritten as [23]

$$\frac{j}{ed} = R(z) + \frac{c\Gamma}{N_g \Delta E} \int_0^{\Delta E} g_m(z, E) S(z, E) \, dE \tag{5.27}$$

The next step is to deduce the spatial dependence of photon density $S(z, E)$. We will derive this in two steps. We first consider the photons of the mth guided mode. It is described by equation (5.25), which can be rewritten as

$$\frac{dS_m}{dt} = \beta R_{sp}(z) + \frac{c\Gamma}{N_g} [g_m(z, E_m) - \alpha_a(z, E_m)] S_m(z, E_m) \tag{5.28}$$

where R_{sp} is the radiative recombination rate of carriers due to spontaneous emissions, E_m corresponds to photon energy of the mth mode and we have replaced S_m/τ_p by $\Gamma \alpha_a/N_g$, where α_a is the loss in the active region [3]. The time derivative dS_m/dt can be regarded as a total derivative and can be expanded as

$$\frac{dS_m^\pm}{dt} = \frac{\partial S_m^\pm}{\partial t} \pm \frac{c}{N_g} \frac{\partial S_m^\pm}{\partial z} \tag{5.29}$$

provided that $S_m^\pm$ varies with z only (i.e. a one-dimensional structure; see subsection 5.4.2). Here, the superscript $\pm$ indicates whether the photons are travelling in a forward or backward direction. A travelling-wave equation can be found by substituting equation (5.29) into equation (5.28). For the spontaneous emission photons the phase information is not important as their major effects are on the intensity of the signals only [22]. Hence it is legitimate and more convenient to work with a travelling-wave equation in terms of intensity for spontaneous emissions, which is given by $I_{sp,m} = c S_{sp,m} E_m/N_g$. Here $S_{sp,m} E_m$ is the total optical energy density of the spontaneous emissions and c/N_g takes care of the number of photons crossing the transverse x–y plane per unit time. Substituting these into equation (5.28) and under a steady-state condition we have

$$\frac{\partial I_{sp,m}^\pm(z, E_m)}{\partial z} = \pm \frac{\beta R_{sp}(z) E_m}{2} \pm g I_{sp,m}^\pm(z, E_m) \tag{5.30}$$

where g is the modal gain coefficient of the amplifier, and we have replaced β by $\beta/2$ to take into account the two possible travelling directions [23]. The first term on the right-hand side of the above equation represents spontaneous emissions generated and coupled to the mth guided mode at z. The second term represents amplified spontaneous emissions travelled to z. It can be shown that if R_{sp} and g_m are independent of z, then equation (5.30) can be solved analytically if the boundary conditions are also known [23]. Furthermore, if β is small and g is relatively small throughout the entire amplifier, the contribution of spontaneous emission $I_{sp,m}^\pm$ to the total photon density S_m will be negligible, i.e. only signal photons will dominate in the interaction between photons and carriers. This is an important point which will arise later in our discussion of using transfer matrices to analyse SLAs.

With $I_{sp,m}$ known, $S(z, E)$ can be found from equation (5.26) in the absence of the signal by $S(z, E) = \Sigma_m S_{sp,m}(z, E)$. To include the signal photons, note that the signal fields satisfy a travelling-wave equation of the form [3, 23]

$$\frac{\partial F^{\pm}(z, E_m)}{\partial z} = \pm \frac{g(z, E)F^{\pm}(z, E)}{2} \mp j\beta_z F^{\pm}(z, E) \qquad (5.31)$$

where β_z is the propagation constant of the field and g is the modal gain coefficient of the amplifier. It should be noted that the unit of $F(z, E)$ is [length]$^{-3/2}$, such that square of the magnitude of $F(z, E)$ represents the photon density of the signal [23]. The solutions of the above equation can be found by imposing the appropriate boundary conditions, which depend on the structure of the SLA. Since $S(z, E)$ now represents the total photon density, the total density of photons in energy bandwidth ΔE (measured from the peak gain photon energy) is the sum of the photon density, the signals and that of spontaneous emissions, which is given in functional form by [23]

$$S(z, E) = [|F^{+}(z, E)|^2 + |F^{-}(z, E)|^2]\delta(E - E_s)\Delta E \\ + 2[I_{sp}^{+}(z, E) + I_{sp}^{-}(z, E)]\delta(E - E_m)\Delta E N_g/cE \qquad (5.32)$$

The factor 2 in front of the spontaneous emitted photons takes into account the two possible polarisations which the spontaneous emissions can take. $I_{sp}^{\pm}$ represents the envelope of the amplified spontaneous emission intensity spectrum across the m possible modes. The delta function $\delta(E - E_m)$ selects that of the signal within ΔE.

Equations (5.27), (5.30) and (5.31) are the key equations in analysing SLAs using travelling-wave equations. The procedure for solving for the gain profile is as follows. We first solve for the signal field distribution at each point along z by using equation (5.31) by assuming a particular form of gain profile initially, imposing suitable boundary conditions like input signal power and facet reflectivities. We then solve equation (5.30) by imposing these boundary conditions and the initial gain profile. Because R_{sp} is z-dependent, the corresponding field solution $F^{\pm}(z, E)$ and the spontaneous emission intensity $I_{sp}^{\pm}(z, E)$ have to be substituted into equation (5.27) in order to obtain a solution for the carrier density and hence the corresponding gain profile g_m by iteration and self-consistency. This is a tedious process, as discovered by Marcuse [3] in his pioneering work. Adams *et al.* [23] simplify this substantially by postulating that instead of working with S_m, which is z-dependent, we can use the photon density averaged along the z-direction of the cavity in solving these equations. Moreover, R_{sp} is assumed to be uniform along the cavity. This avoids the need for excessive iterative computation to solve for the field at each point along z [28]. With these assumptions, equation (5.32) is replaced by [23]

$$S(z, E) \approx S_{ave} = [S_{1ave}\delta(E - E_r)\beta(E_r) + S_{2ave}\delta(E - E_s)]\Delta E \qquad (5.33a)$$

where

$$S_{1ave} = \\ \left\{ \frac{(\exp(gL) - 1)[(1 - R_2)(1 + R_1\exp(gL)) + (1 - R_1)(1 + R_2\exp(gL))]}{gL[1 - R_1R_2\exp(2gL)]} - 2 \right\} \frac{R_{sp}N_g}{gc}$$

$$(5.33b)$$

$$S_{2\text{ave}} =$$

$$\left\{ \frac{(1 - R_1)(\exp(gL) - 1)[1 + R_2 \exp(gL)]}{(1 - \sqrt{R_1 R_2} \exp(gL))^2 + 4\sqrt{R_1 R_2} \exp(gL) \sin^2((2\pi N L/hc)(E - E_r))} \right\} \frac{P_{\text{in}} N_g}{EWdcgL}$$

$$(5.33c)$$

where N is the effective refractive index, and $S_{1\text{ave}}$ and $S_{2\text{ave}}$ represent the average spontaneous emission photon density and the signal photon density along the cavity, respectively. The average spontaneous emission photons can be found by solving equation (5.30) alone (i.e. considering that there is no input signal to the amplifier) with R_{sp} independent of z and imposing appropriate boundary conditions. $\beta(E_r)$ takes into account the proportion of spontaneous emission which is actually coupled to the cavity modes. This gives a simpler way to solve the equations without involving tedious computations (since the number of equations to be solved reduces to three instead of *three for every point* along z) but it has surprisingly good accuracy [28]. In this manner, an average gain coefficient g_{ave} can be obtained (instead of a number of gain coefficients $g_m(z)$ along z which form the gain profile), and this can be substituted subsequently into equation (5.12) in subsection 5.2.1 to find the amplifier gain for a particular input power. By proceeding to calculate the amplifier gain for different input optical power levels, the gain saturation characteristics can also be derived. Later in this chapter we will show some results calculated by the modified method proposed by Adams *et al.* [37] as described above. Before that, we will look at an alternative method to solve for the travelling-wave equations, equations (5.27), (5.30) and (5.31). This involves a special technique known as the transfer matrix method (TMM) [4, 5, 70].

5.4 ANALYSIS OF SEMICONDUCTOR LASER AMPLIFIERS USING TRANSFER MATRICES

5.4.1 A brief review of matrix methods

We saw in the previous section that if we have to analyse SLAs using travelling-wave equations, which have taken into account the spatial dependence of the fields and carrier density (hence the material gain profile), three unknown quantities have to be found from equations (5.26), (5.30) and (5.31) *for each value of z*. These quantities are signal fields $F^{\pm}$, spontaneous emissions intensity $I_{\text{sp}}^{\pm}$, and the carrier density n (for simplicity we have dropped the functional dependence of these quantities on z and E). Hence, as discussed in subsection 5.3.2, the solution will require a tedious process of numerical iteration. A good example to illustrate the complexity involved in analysing SLAs with travelling-wave equations is to consider SLAs using distributed feedback (DFB) structures [38, 39]. It is well known that to find the longitudinal field $F^{\pm}(z)$ and the longitudinal propagation constant β_z, in DFB laser structures having only a knowledge of transverse modal fields is not sufficient. This is in contrast to the solution of the longitudinal propagation constant in simple buried heterostructures, as discussed in Chapter 4, where β_z can be found by equation (4.26) once the effective index has been determined by transverse modal field analysis. A coupled wave equation has to be solved numerically [33]. In addition, the longitudinal fields have to satisfy equation (5.31) in an amplifier. Hence, over-all analysis of the amplifier characteristics becomes much more complicated and time consuming.

One possible method of simplifying the analysis procedure, whilst increasing its flexibility and robustness, is to employ matrix methods. Matrices have been used in many engineering problems of an intensive numerical nature such as solving transverse modal fields in arbitrarily shaped dielectric waveguides [40, 41]. In microwave engineering, matrix methods have been used to solve for electric and magnetic fields in waveguides and microwave devices [42–44]. The advantage of using matrices is that the algorithm involved in solving the fields is identical for all types of structures with different boundary conditions. Thus the method of analysis becomes more flexible. However, because matrix methods are purely numerical in nature, they do not allow us to detect whether a closed-form analytical result exists for the structure. Such analytical solutions can be useful as they offer a quick way to estimate how the solutions will vary with various parameters.

In all matrix methods the structure under analysis will be divided into a number of sections, which can be either fairly large (e.g. in the transmission-line matrix (TLM) method [42]) or as few as two or three (e.g. in scattering matrices [10]). Each section is modelled by an N-port network, with fields travelling in both directions at each port (see Figure 5.4). The value of N will be determined by the number of interfaces joining neighbouring sections. For example, for one-dimensional structures, $N = 2$. There is a relation between the fields at each port for each section. At the ith section in a one-dimensional structure, this relation can be represented by the following matrix equation:

$$F_{1,i} = A_i \cdot F_{2,i} \tag{5.34}$$

where $F_{1,i}$ and $F_{2,i}$ are the column vectors representing the fields at port 1 and port 2, respectively, and A_i is the matrix the coefficients of which will determine the relationship between the fields $F_{1,i}$ and those in $F_{2,i}$. If the structure under analysis is one-dimensional, then at each port there can only be two fields, travelling into and out of the port, respectively [5]. Using equation (5.34), $F_{1,i}$

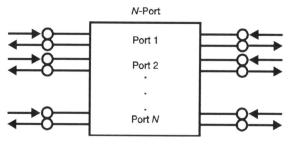

Figure 5.4 Representation of a section in a structure analysed by matrix methods with an N-port system

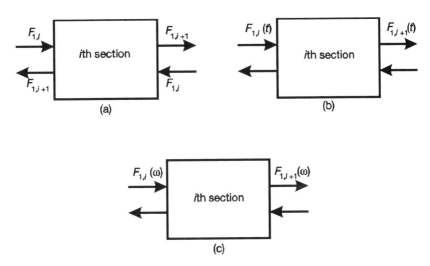

Figure 5.5 Various matrix methods used to analyse electromagnetic fields in different structures. (a) Scattering matrix, (b) TLM and (c) TMM

and $F_{2,i}$ can be found for each of the ith sections, noting that $F_{2,i} = F_{1,i+1}$ when the entire structure is represented by cascading such two-port sections, each with a matrix A. Then the relation between $F_{1,i}$ and $F_{1,M+1}$ can be found via the following equation:

$$F_{1,i} = A \cdot F_{1,M+1} \tag{5.35}$$

for a structure with M sections. The matrix A is determined by the products of the matrix A_i of each section, that is

$$A = \prod_{i=1}^{M} A_i \tag{5.36}$$

Since either $F_{1,i}$ or $F_{1,M+1}$ is usually determined by the boundary conditions imposed on the structure (e.g. input power to the structure), the fields at each interface joining the ports can be found from equation (5.34), provided that the coefficients in A_i are known. Usually, the coefficients of A_i are determined by the structural and material characteristics of that section and can be determined for each section. If the number of sections used is sufficiently large, the continuous field profile within the structure can also be determined accurately.

The exact forms of $F_{1,i}$ and $F_{1,i+1}$ are different for different matrix methods. Some are shown in Figure 5.5. In scattering matrix analysis, $F_{1,i}$ will represent all the fields travelling into the port, whilst $F_{i,i+1}$ represent all the fields travelling out of the port [10]. Notice that strictly speaking the fields travelling out of the ith section should be denoted as $F_{i,i\pm1}$ because one will enter the $(i + 1)$th section, whilst the other field will enter the $(i - 1)$th section. In the transmission line matrix (TLM) analysis, $F_{1,i}$ represents all the fields at one port, whereas $F_{i,i+1}$ represents those at the other port [42]. The difference between TLM and the Transfer Matrix Method (TMM) is that the former analysis will be performed in the time domain, whereas the latter one is in the frequency domain.

Equations (5.34)–(5.36) have been derived under the assumption that the structure under analysis is one-dimensional. In practice, all of these matrix methods can be extended to analyse three-dimensional structures. However, the algorithm involved is not only more complex, but is also unstable numerically [45]. In addition, as we shall see later, a three-dimensional structure can be modelled by an equivalent one-dimensional structure by dividing the structure appropriately and if some pre-analysis has been performed beforehand. This works exceptionally well in laser amplifiers as they are essentially two-port devices with one input and one output.

Both TLM [45] and TMM [5] have been applied to analyse semiconductor laser devices. The difference in the domains of analysis of these two methods also accounts for their differences in applications. TLM works exceptionally well for analysis of the transient responses of both semiconductor lasers and laser amplifiers [35, 45, 46], although it can also be used to analyse steady-state performance relatively easily too. However, it is difficult to use TLM to analyse the noise performance of laser devices, e.g. the spectral linewidth in semiconductor lasers, and the noise figure and phase noise in laser amplifiers. This is because most of the noise-related phenomena are time-averaged stochastic processes [47]. Using TLM in the time domain to analyse noise processes will involve sampling over a very long period of time. Although this is feasible for very simple analysis (e.g. see Lowery [48]), TLM is not suitable in general to analyse the noise characteristics in semiconductor laser devices.

On the other hand, because analyses by TMM are performed in the frequency domain, both the steady-state and noise characteristics (e.g. spectral linewidth) for semiconductor laser devices can be obtained accurately and efficiently [5, 38, 49]. However, the transient responses of these laser devices cannot be determined efficiently as a very wide frequency range will be required for accurate solutions in the analysis. In this book we are primarily interested in the steady-state operations of SLAs in optical communications. Hence, the TMM will be more suitable for our foregoing studies of their gain, saturation and noise characteristics. In the following subsection, we examine how to implement the TMM using equations (5.34)–(5.36) to analyse the longitudinal travelling fields in SLAs.

5.4.2 Analysis of longitudinal travelling fields in SLAs using the TMM

We saw in subsection 5.3.2 that it is vital to find the longitudinal field distribution within a SLA in order to determine the material gain profile in the amplifier (equation (5.27)) and subsequently the overall amplifier character-istics. An efficient and robust technique of analysis which can yield this distribution will be ideal. The matrix formalism outlined in the previous discussion will be used here to develop such a technique. Most SLAs are buried

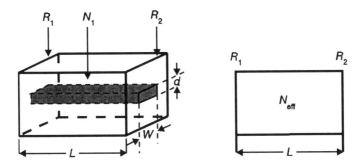

Figure 5.6   Modelling of a buried heterostructure by a one-dimensional waveguide

heterostructures [46]. For these structures, as we have examined in the previous chapter, the active region that provides gain can be represented by a rectangular dielectric waveguide of refractive index N_1 embedded in a material with a different refractive index N_2. Using the transverse modal field analysis outlined in the previous chapter, we can deduce the effective group refractive index N_{eff} for the propagating signal. The x–y dependence of the fields has then been eliminated. Recall that the total solution of the electric or magnetic fields in such a dielectric waveguide is given by [10]

$$E_i(x, y, z) = \Phi_m(x, y)F(z), \quad i = x, y, z \tag{5.37}$$

Knowledge of $\Phi_m(x, y)$ allows us to use a simple one-dimensional waveguide to represent the true three-dimensional structure. This is illustrated in Figure 5.6. We will use this one-dimensional structure in the foregoing discussion on the implementation of the TMM in analysing SLAs. Before we examine how to use the TMM to analyse the longitudinal fields in the SLA, the reader should notice that our preceding discussion was with a BH SLA. This type of amplifier can indeed be modelled very easily by a rectangular dielectric waveguide [50]. However, modern SLAs involve more complex structures like ridge or rib waveguides [4]. In these cases the analyses of the transverse modal fields are indeed more complicated than outlined in Chapter 4 [51]. However, once the field solutions $\Phi_m(x, y)$ and the corresponding effective refractive index N_{eff} are calculated from the transverse modal field analysis, we can use equation (5.37) to analyse $F(z)$ by using a one-dimensional waveguide model with a refractive index of N_{eff}, as illustrated in Figure 5.6. Therefore the general implications of using equation (5.37) with the effective index remains. But it should be noted that the form of equation (5.37) assumes that $\Phi_m(x, y)$ and $F(z)$ are separable. This is true for index guided structures as the propagation constant β_z is independent of x, y or z [18]. For a gain guided structure (e.g. a stripe geometry laser [52]), this assumption may not hold [53].

We consider an amplifier that can be modelled as a one-dimensional waveguide along z (Figure 5.7). The structure under analysis is a general one in which we can divide the waveguide along z into a number of smaller cavities, each with an effective index N_{ei}. Such a general description will be useful in

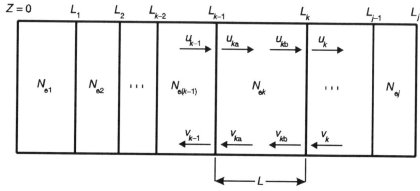

Figure 5.7 A multi-cavity equivalent of one-dimensional waveguide used in TMM analysis

analysing more complex structures like two-section amplifiers [54]. We will first look at the kth cavity with an effective index N_{ek} and length $L = L_k - L_{k-1}$. As discussed in the previous subsection there can only be two fields entering the one-dimensional kth cavity only, which are denoted as u_{k-1} and v_k, respectively (see Figure 5.7). Similarly, there are only two fields leaving the waveguide, which are denoted as u_k and v_{k-1}, respectively. We have shown in the previous section that a relationship exists between these fields. Using equation (5.34) under the transfer matrix formalism, the following matrix equation can be derived:

$$\begin{bmatrix} u_k \\ v_k \end{bmatrix} = \begin{bmatrix} a_{11}^k & a_{12}^k \\ a_{21}^k & a_{22}^k \end{bmatrix} \begin{bmatrix} u_{k+1} \\ v_{k+1} \end{bmatrix} \tag{5.38}$$

The matrix on the right-hand side of equation (5.38) is the transfer matrix A_k of the kth cavity. Our next step is to derive the coefficient of A_k. A careful examination of the kth cavity in Figure 5.7 reveals that it is made up of (i) a refractive index step between $N_{e(k-1)}$ and N_{ek} at $z = L_{k-1}$; (ii) a uniform one-dimensional waveguide along z with length L and an effective index N_{ek}; and (iii) a refractive index step between N_{ek} and $N_{e(k-1)}$ at $z = L_k$. It is well known that any refractive index steps along the propagation direction of the fields will induce reflections. Therefore a relationship exists between the fields u_k, v_k and those *immediately after* the interface, i.e u_{ka}, v_{ka}, at $z = L_{k-1}$ [10]. Hence we can express u_k and v_k as [5, 49]

$$\begin{bmatrix} u_k \\ v_k \end{bmatrix} = \frac{1}{t_{k-1}} \begin{bmatrix} 1 & -r_{k-1} \\ -r_{k-1} & 1 \end{bmatrix} \begin{bmatrix} u_{ka} \\ v_{ka} \end{bmatrix} \tag{5.39}$$

where

$$r_{k-1} = \frac{N_{e(k-1)} - N_{ek}}{N_{e(k-1)} + N_{ek}} \tag{5.40}$$

In equation (5.40) r_{k-1} is the *amplitude* reflectivity of the interface at $z = L_{k-1}$.

The quantity t_{k-1} is the field *amplitude* transmittivity of the same interface, and is related to r_{k-1} by [10]

$$t_{k-1}^2 = 1 - r_{k-1}^2 \tag{5.41}$$

Similarly, at the interface $z = L_k$, one can relate the fields u_{k+1} and v_{k+1} with those *immediately before* the interface by

$$\begin{bmatrix} u_{kb} \\ v_{kb} \end{bmatrix} = \frac{1}{t_k} \begin{bmatrix} 1 & r_k \\ r_k & 1 \end{bmatrix} \begin{bmatrix} u_{k+1} \\ v_{k+1} \end{bmatrix} \tag{5.42}$$

where

$$r_k = \frac{N_{e(k+1)} - N_{ek}}{N_{e(k+1)} + N_{ek}} \tag{5.43}$$

In equation (5.43) r_k is the amplitude reflectivity of the interface at $z = L_k$. The quantity t_k is therefore the amplitude transmittivity of the same interface which is given by

$$t_k^2 = 1 - r_k^2 \tag{5.44}$$

We have now established a relationship between (u_k, v_k) and (u_{ka}, v_{ka}) (see equation (5.39)), and a similar relationship between (u_{k+1}, v_{k+1}) and (u_{kb}, v_{kb}). If a relation between the internal fields (u_{ka}, v_{ka}) and (u_{kb}, v_{kb}) can be established, then the coefficients in the matrix A_k can be found. This can be achieved by noting that the rest of the kth cavity is a uniform one-dimensional waveguide. Since we have been analysing structures with index guiding, we can derive the relation between (u_{ka}, v_{ka}) and (u_{kb}, v_{kb}) by solving the one-dimensional wave equation (i.e. the travelling-wave equation, equation (5.31)) [55, 57]:

$$\frac{d^2 F(z)}{dz^2} = -\frac{\omega^2}{c^2} N_{ek}^2 F(z) \tag{5.45}$$

where $F(z)$ can be found to have a general solution of the form

$$F(z) = u \cdot \exp(-j\beta_z z) + v \cdot \exp(j\beta_z z) \tag{5.46}$$

Then it can be seen immediately that

$$u_{ka} = u_{kb} \cdot \exp(j\beta_z L) \tag{5.47a}$$

$$v_{ka} = v_{kb} \cdot \exp(j\beta_z L) \tag{5.47b}$$

Hence, equation (5.38) can be rewritten as

$$\begin{bmatrix} u_k \\ v_k \end{bmatrix} = \frac{1}{t_k t_{k-1}} \begin{bmatrix} 1 & -r_{k-1} \\ -r_{k-1} & 1 \end{bmatrix} \begin{bmatrix} \exp(j\beta_z L) & 0 \\ 0 & \exp(-j\beta_z L) \end{bmatrix} \begin{bmatrix} 1 & r_k \\ r_k & 1 \end{bmatrix} \begin{bmatrix} u_{k+1} \\ v_{k+1} \end{bmatrix} \tag{5.48}$$

β_z can be found easily since the effective index is known (see equation (4.26) in Chapter 4), that is

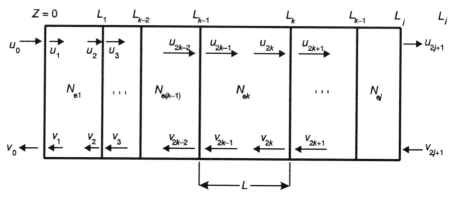

Figure 5.8   Formalising the transfer matrix method to analyse a multi-cavity structure

$$\beta_z = \frac{\omega N_{ek}}{c} \tag{5.49}$$

Hence all the coefficients in the matrices of equation (5.48) are known once the structural parameters N_{ek}, $N_{e(k+1)}$ and L are known. Then if u_{k+1} and v_{k+1} are known, we can easily "transfer" them to form the quantities u_k and v_k by using equation (5.48). The above derivation holds for one cavity only. To formalise the analysis we should consider all K-sections shown in Figure 5.7. In doing so we have to redefine the internal fields as shown in Figure 5.8. The relationship between the fields at the interface are given by equations (5.39) and (5.42), respectively. The fields within the waveguide are related by equations (5.47a) and (5.47b). In this manner u_{2k+1} and v_{2k+1} are related to u_0 and v_0 via the following matrix equation:

$$\begin{bmatrix} u_0 \\ v_0 \end{bmatrix} = \begin{bmatrix} a_{11} & a_{12} \\ a_{21} & a_{22} \end{bmatrix} \begin{bmatrix} u_{2k+1} \\ v_{2k+1} \end{bmatrix} \tag{5.50}$$

where

$$A = \begin{bmatrix} a_{11} & a_{12} \\ a_{21} & a_{22} \end{bmatrix} = A_{01} \cdot A_{12} \cdot A_{23} \cdots A_{2k,2k+1} \tag{5.51}$$

The subscripts $_{ij}$ (i.e. A_{ij}) in equation (5.51) represent the matrix which relates the electrical fields u_i, v_i to those of u_{i+1}, v_{j+1}. Since the coefficients of the above transfer matrix can be derived by using equations (5.39), (5.42), (5.47a), (5.47b) and (5.48) from the structural parameters, once the fields at $z = 0$ or $Z = L_k$ are known, then all the fields u_i and v_i along the structure can be calculated including the fields at $z = 0$ (if the fields at $z = L_k$ are known) or the fields at $z = L_k$ (if the fields at $z = 0$ are known), and hence the signal gain of the amplifier. The reader may notice from equations (5.47a) and (5.47b) that in solving the fields in the one-dimensional waveguide, we have so far neglected the fact that there is optical gain in the structure (since it is a laser amplifier). To introduce this fact into the analysis, we can follow the procedures outlined in Section 4.3 in Chapter 4 by introducing an imaginary part to the longitudinal propagation constant. In this case

$$\beta_z \rightarrow \beta_z + jg/2 \tag{5.52}$$

where g is the modal gain coefficient of the active region. Then for the kth cavity, equation (5.48) can be rewritten as

$$
\begin{bmatrix} u_k \\ v_k \end{bmatrix} =
$$

$$
\frac{1}{t_k t_{k-1}} \begin{bmatrix} 1 & -r_{k-1} \\ -r_{k-1} & 1 \end{bmatrix} \begin{bmatrix} \exp(gL/2 + j\beta_z L) & 0 \\ 0 & \exp(gL/2 - j\beta_z L) \end{bmatrix} \begin{bmatrix} 1 & r_k \\ r_k & 1 \end{bmatrix} \begin{bmatrix} u_{k+1} \\ v_{k+1} \end{bmatrix}
\tag{5.53}
$$

Notice that equation (5.53) has been derived under the assumption that *g is uniform along L*. In addition, if β_z has taken the form of equation (5.52), the effective indices of each cavity become complex according to equation (5.49). Then the corresponding reflectivities and transmittivities should also become complex. However, as discussed in Section 4.3, the imaginary part in equation (5.52) is much smaller compared with the real part. Hence the imaginary parts of the effective indices will also be very small. Thus we can continue to calculate the reflectivities and transmittivities at interfaces joining the cavities using the real parts of the effective indices only without significant loss in accuracy [5]. In the following subsection we examine how to modify the analysis described here to consider the situation when *g* is not uniform along *L*.

5.4.3 *Analysis of SLAs with a non-uniform gain profile using the TMM*

To simplify the foregoing discussion we consider SLAs with one cavity only ($K = 1$), as shown in Figure 5.9. The analysis described here can be easily extended to SLAs having more than one cavity following the procedures illustrated in Figure 5.8 and equations (5.50) and (5.51). As illustrated in Figure 5.9, the amplifier is modelled by a one-dimensional cavity with effective

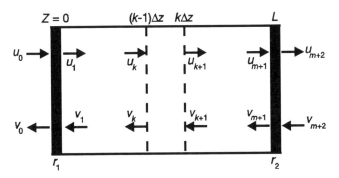

Figure 5.9 The model for analysis of a SLA using the TMM by taking into account the non-uniform gain profile

index N_{eff} and length L. We further divide the cavity into M sections, each of length Δz. The modal gain coefficient of the kth section is given by $g(z = k\Delta z) = g_k$, and can be found from equations (4.49) and (4.53) if the material gain coefficient g_{m} is known. For simple BH SLAs, g_k is related to $g_{\text{m}}(z = k\Delta z)$ by [59]

$$g_k = \Gamma(g_{\text{m}k} - \alpha_{\text{a}k}) - (1 - \Gamma)\alpha_{\text{c}k} \tag{5.54}$$

where we have used $g_{\text{m}k} = g_{\text{m}}(z = k\Delta z)$, and similarly $\alpha_{\text{a}k}$ and $\alpha_{\text{c}k}$ are the losses in the active region and in the claddings, respectively. As discussed in subsection 5.3.2, the material gain coefficient $g_{\text{m}k}$ and the amplitude of the electric fields are not unrelated. This is because $g_{\text{m}k}$ is determined by the carrier density $n(z = k\Delta z) = n_k$. Under a linear model [13]

$$g_{\text{m}k} = A(n_k - n_0) \tag{5.55}$$

where n_0 is the transparency carrier density of the material (see subsection 3.2.1 in Chapter 3). Notice that n_0 depends on the band structure of the material only, and is assumed to be independent of z. In addition, we have assumed that the signal wavelength is matched to that of the peak of the material gain spectrum, so that $g_{\text{m}k}$ is the peak gain coefficient of the material. On the other hand, the carrier density n_k is related to the optical fields at $z = k\Delta z$ via the rate equation (equation (5.27)):

$$\frac{j}{ed} = R_k + \frac{c\Gamma}{N_{\text{g}}} g_{\text{m}k} S_k \tag{5.56}$$

where R_k is the recombination rate of the carrier at $z = k\Delta z$ and $S_k = S(z = k\Delta z)$ is the photon density at $z = k\Delta z$. In equation (5.56) we have assumed that the current density j is uniform across the active region and the contribution of spontaneous emissions are neglected by assuming that $g_{\text{m}k}$ is small. Thus S_k consists of photons due to the input signal. If we normalise the fields u_{k+1} and v_{k+1} such that the square of their magnitudes represents the intensity of the optical signals travelling into and out of the kth section, then we have [37]

$$S_k = [|u_{k+1}|^2 + |v_{k+1}|^2] \cdot \frac{N_{\text{g}}}{cE} \tag{5.57}$$

where E is the energy of the signal photons. Notice that in deriving equation (5.57), the coupling of any spontaneous emission $I_{\text{sp}}^{\pm}$ to the guided mode of signals u_{k+1} and v_{k+1} has been neglected (i.e. $\beta \approx 0$ in equation (5.33)). As we shall see later in subsection 5.4.4, the contribution of spontaneous emissions to the amplifier characteristics is not important in analysing SLAs *provided that the bias of the amplifier is not large and $g_{\text{m}k}$ is small for all k's*. Usually this implies that the amplifier is biased below the oscillation threshold [37]. Since the performance of SLAs biased well below the oscillation threshold is of most interest, the above assumption will be valid for most of our foregoing analysis. Substituting equation (5.57) into equation (5.56) yields the following equation:

$$\frac{j}{ed} = R_k + \frac{\Gamma g_{\text{m}k}}{E} [|u_{k+1}|^2 + |v_{k+1}|^2] \tag{5.58}$$

By solving equation (5.58) for g_{mk} and hence n_k by using equation (5.55), we can deduce the coefficients of the matrix A_k, which relates (u_k, v_k) and (u_{k+1}, v_{k+1}):

$$\begin{bmatrix} u_k \\ v_k \end{bmatrix} =$$

$$\begin{bmatrix} \exp(g_k \Delta z/2 + j\beta_z \Delta z) & 0 \\ 0 & \exp(g_k \Delta z/2 - j\beta_z \Delta z) \end{bmatrix} \begin{bmatrix} u_{k+1} \\ v_{k+1} \end{bmatrix} \tag{5.59}$$

Equations (5.58) and (5.59) are the core equations in our analysis of SLAs using the TMM. By solving them, we can find (i) the field, (ii) the material gain coefficient and (iii) the carrier density at $z = k\Delta z$ along the amplifier. Furthermore, the format of equation (5.59) suggests that the computational algorithm (see the next subsection) will converge unconditionally, unlike solving the travelling-wave equations by iteration and self-consistency [3]. Any facet reflectivities can also be accounted for with equations (5.39) and (5.42) (Figure 5.9), such that

$$\begin{bmatrix} u_{M+1} \\ v_{M+1} \end{bmatrix} = \frac{1}{t_2} \begin{bmatrix} 1 & r_2 \\ r_2 & 1 \end{bmatrix} \begin{bmatrix} u_{M+2} \\ v_{M+2} \end{bmatrix} \tag{5.60}$$

and

$$\begin{bmatrix} u_0 \\ v_0 \end{bmatrix} = \frac{1}{t_1} \begin{bmatrix} 1 & -r_1 \\ -r_1 & 1 \end{bmatrix} \begin{bmatrix} u_1 \\ v_1 \end{bmatrix} \tag{5.61}$$

Thus by solving for (u_0, v_0) for given values of (u_{M+2}, v_{M+2}), we can calculate the amplifier gain by

$$G = \left| \frac{u_{M+2}}{u_0} \right|^2 \tag{5.62}$$

By calculating G for different levels of output optical power P_{out}, we can derive the variation of G against P_{out} and hence deduce the saturation output power.

5.4.4 Computational considerations

In this subsection the technique described above will be used to analyse a simple Fabry–Perot amplifier (FPA), which will be compared with the analysis proposed by Adams *et al.* [37]. The computation procedure with the TMM is as follows:

(i) We start with an initial value of P_{out}. This quantity is related to u_{M+2} and v_{M+2} by

$$P_{out} = |u_{M+2}|^2 Wd \tag{5.63a}$$

and

$$v_{M+2} = 0 \tag{5.63b}$$

(ii) Compute u_{M+1} and v_{M+1} by using equation (5.60).

(iii) Compute the fields u_M and v_M using equation (5.59) with $k = M$.

The simplest way is to express equation (5.58) in terms of n_k by using equation (5.55). The recombination rate R_k for undoped materials for which Auger recombinations are not important is given simply by [37]

$$R_k = Bn_k^2 \tag{5.64}$$

In long-wavelength semiconductor laser devices we have to take into account the Auger recombinations; that is, the following R_k expression should be considered [71]:

$$R_k = A_{nr}n_k + Bn_k^2 + Cn_k^3 \tag{5.65}$$

By substituting equation (5.64) or (5.65) and equation (5.55) into equation (5.58), we can calculate n_k and hence g_{mk}. Then g_k can be calculated from equation (5.54) and the coefficients of the matrix A_k in equation (5.59) can be formed. Finally, u_k and v_k can be calculated. The above steps provide us with three vital pieces of information: the carrier density n_k, the material gain coefficient g_m and finally the fields u_k and v_k. If M is large enough, we can derive the material gain profile and the longitudinal field distribution along z. This is not possible with the method proposed by Adams *et al.* [37].

   (iv) Repeat step (iii) until u_1 and v_1 are calculated at $M = 1$.

   (v) Calculate the input fields u_0 and v_0 by using equation (5.61) and hence the input optical power P_{in} using the following equation:

$$P_{in} = |u_0|^2 \tag{5.66}$$

In the absence of spontaneous emission coupled to the signal and the boundary condition of $v_{M+2} = 0$, we can show easily that $v_0 = 0$.

   (vi) The amplifier gain is calculated by using equation (5.62).

   (vii) Repeat steps (i)–(vi) for a different value of P_{out} until all the values of P_{out} in the range of interest have been covered.

To compare the above technique with that proposed by Adams *et al.* [37], a simple FPA with the structural parameters listed in Table 5.2 has been analysed. Auger recombinations have been neglected for simplicity. We first compare the speed of computation between the two techniques. For $M = 300$ sections, the TMM uses approximately 0.8 s on an IBM3090 mainframe computer to calculate 50 values of G in the output power range of −60 dBm to 0 dBm for a fixed bias current. For a similar set of calculations using the technique suggested by Adams *et al.* [37], it takes approximately 1.1 s. Though the difference is not much, it should be pointed out that there are some points that do not converge using the latter method, especially when G starts to saturate. In fact, it took several days for us to properly adjust the algorithm suggested by Adams *et al.* [37] such that the initial trial value could lead to speedy computations with a sufficient number of converged results. This critical importance of the initial value has also been discussed recently by Zaglanakis and Seeds [58]. On the other hand, the algorithm based on the

Table 5.2 SLA parameters used in calculations [16, 37]

Length of the amplifier	$L = 500\,\mu m$
Width of the active region	$W = 1.5\,\mu m$
Thickness of the active region	$d = 0.2\,\mu m$
Optical confinement factor	$\Gamma = 0.5$
Total loss coefficient	$\alpha = 25\,cm^{-1}$
Group effective index	$N_g = 3.5$
Reflection coefficients	$R_1 = R_2 = 0.36\%$
Differential gain coefficient	$A = 4.63 \times 10^{-16}\,cm^2$
Radiative recombination coefficient	$B = 10^{-10}\,cm^3\,s^{-1}$
Spontaneous emission coefficient	$\beta = 10^{-4}$

TMM does not need a lot of adjustment and fine-tuning (we needed only half an hour to write a successful working program for this analysis). Although the above comparisons are difficult to quantify and may depend on the experience and skills of the programmer, it clearly illustrates the ease of implementation and robustness of the TMM algorithm.

Figure 5.10 shows the results of the calculated value of the amplifier gain G versus output optical power P_{out} using both the averaged photon density (AVPD) approximation [37] and the transfer matrix method described above. Both $\beta = 0$ and $\beta = 10^{-4}$ have been considered in the AVPD calculations. As discussed previously, the TMM approach will consider $\beta = 0$. Results shown

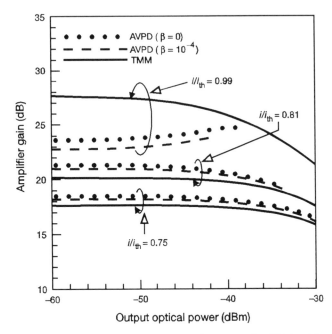

Figure 5.10 Amplifier gain versus output optical power of a FPA with parameters listed in Table 5.2 calculated by the TMM and averaged photon density approximation

are for three particular pumping levels of $i/i_{th} = 0.75$, 0.81 and 0.99, respectively, where i_{th} is the lasing threshold current for the structure and i is the bias or injection current. There are two important points which can be observed from the diagram. First, the effect of β on the calculated results using the AVPD approximation is not significant until the structure is approaching lasing. This was also discovered by Adams et al. [37] in their studies. Hence for normal operation of the SLA, performing the analysis by taking $\beta = 0$ will be accurate enough (unless the SLA is biased close to the lasing threshold, as in the optical switching application discussed in subsection 3.5.3 in Chapter 3). Secondly, the differences between results calculated by the TMM are not significant (around 1 dB) when the SLA is biased below the threshold. A huge difference is observed when the SLA is driven close to oscillation. Thus, compared with the AVPD approach, the results generated by the TMM are in close agreement with those predicted by the AVPD for low to moderate pumping. However, the differences between the two methods at a high injection level are significant (even the shapes of the curves look different). The accuracy of both of the methods can only be assessed by comparison with experimental results. A more detailed examination of this will be given in Chapter 7.

To illustrate the benefits of using the TMM, a profile of the carrier density n along z and the normalised longitudinal field distribution have been calculated as shown in Figures 5.11 and 5.12, respectively. The distribution for two output powers at $i/i_{th} = 0.81$ are shown. In Figure 5.11 it can be seen that when the output power increases, the carrier density distribution becomes more uniform compared with the lower power level. Thus the AVPD approximation may not be accurate at high output power levels (since it assumes a uniform distribution for the carrier density). This will be explored again in Chapter 7. The total fields (the two curves at the top) and the corresponding components of u and v are plotted in Figure 5.12 with respect to z, at $i/i_{th} = 0.81$ for two different output power levels. Interestingly, we did not obtain a more uniform curve like Adams et al. [37] obtained in their analysis. This is because the latter analysis was performed when the signal wavelength is matched to the cavity resonance of the FPA. From this, they [37] justified their AVPD approach. In the present analysis, where the signal wavelength is not matched to the cavity resonance, a strong non-uniform field distribution is observed. Thus in using the AVPD approach we must be careful to note that the assumptions are closest to reality *only when the signal wavelength is matched with the FP cavity resonance*. This illustrates the clear necessity of taking into account the non-uniform field and hence the photon distribution along z in analysing SLAs.

5.5 AN EQUIVALENT CIRCUIT MODEL FOR SLAs

The use of the TMM in the analysis of SLAs has been illustrated in the previous section. In this section based on the TMM we derive an equivalent circuit to model SLAs [4, 70, 71] and this equivalent circuit model will be used in Chapter 6 to deal with the noise performance of SLAs.

As discussed in subsection 5.4.1, matrix methods were originally applied to

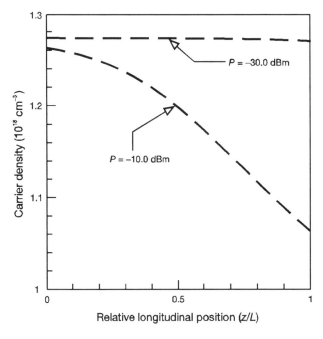

Figure 5.11 Carrier density distribution along the z-direction for the FPA analysed in Figure 5.10

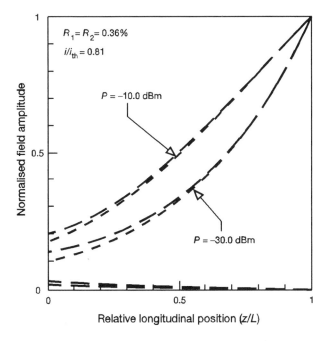

Figure 5.12 Longitudinal field profile along the z-direction for the FPA analysed in Figure 5.10

microwave engineering problems. They have also been applied to semi-conductor laser devices because both methods are concerned with guided wave phenomena. Thus we can actually construct an analogy between the semiconductor laser problem and an equivalent problem in microwave engineering. Such an analogy has been widely used to help engineers to understand and analyse semiconductor laser devices more efficiently, because we are now dealing with the familiar electronic circuit terminology rather than more complex laser physics phenomena. This analogy is called *a model*. Some of the models used by researchers to analyse semiconductor lasers include an equivalent circuit to analyse the electrical properties of laser diodes [59], the negative resistance oscillator model to analyse quantum mechanical fluctuations in lasers [60] and subsequently their spectral linewidth [49], and an equivalent transmission line mesh to analyse laser devices with the TLM [35, 45].

We will develop a new equivalent circuit model for the SLA based on the negative resistance approach [4, 70]. In the earlier works [49, 60], a semi-conductor laser was modelled by a negative resistance oscillator. The negative resistance is a direct analogy to the optical gain in the active region, because both release power rather than absorbing it as in ordinary electrical resistance and passive optical components [60]. However, we cannot directly use this equivalent circuit to analyse SLAs because (i) SLAs are not oscillators and (ii) the above-mentioned equivalent circuit modelled the laser diode as a bulk, i.e. it uses a uniform material gain profile. As mentioned in the previous sections, a uniform gain profile can be used to analyse lasers with sufficient accuracy [61, 62], but will not be good enough for SLAs. An entirely new equivalent circuit must therefore be derived for the analysis of SLAs.

Since a negative resistance Z_k has been used to model the optical gain in the active region in the negative resistance oscillator analysis of semiconductor lasers, it will also be used to represent the modal gain coefficient g_k at the kth section in the one-dimensional waveguide used by the TMM (Figure 5.9). Note that instead of modelling the optical gain of the bulk active region by a single negative resistance, a single negative resistance will be used to represent a section of the amplifier under analysis only. Each section of the amplifier not only provides the gain in the field amplitude, it also introduces a phase delay due to propagation of the $\exp(j\beta_z \Delta z)$ term in A_k (see equation (5.59)). We can represent this by a piece of lossless transmission line of length Δz. The characteristic impedance of the line Z_0 is made such that the propagation constant in the line is identical to that in the amplifier. The equivalent circuit for a section of the amplifier under analysis is therefore derived. This is illustrated in Figure 5.13, which consists of two elements: a lumped negative resistance Z_k and a piece of lossless transmission line of characteristic impedance Z_0 and length Δz. We consider the voltages travelling along the $+z$-direction in this equivalent circuit. By simple circuit analysis, at the steady state we have

$$V = \left(\frac{Z_0}{Z_k + Z_0} \right) \cdot V_k \tag{5.67}$$

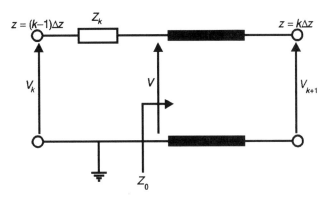

Figure 5.13 An equivalent circuit for fields travelling in the z-direction in a section of a SLA

$$V_{k+1} = V \cdot \exp(-j\beta_z \Delta z) \qquad (5.68)$$

where the frequency dependence on the voltages is assumed implicit. Hence:

$$\frac{V_{k+1}}{V_k} = \left(\frac{Z_0}{Z_k + Z_0} \right) \cdot \exp(-j\beta_z \Delta z) \qquad (5.69)$$

Comparing this with equation (5.59), it can be seen that

$$\frac{V_{k+1}}{V_k} \equiv \frac{u_{k+1}}{u_k} = \exp\left(\frac{g_k \Delta z}{2} \right) \exp(-j\beta_z \Delta z) \qquad (5.70)$$

Therefore the following analogy can be made:

$$\left(\frac{Z_0}{Z_k + Z_0} \right) = \exp\left(\frac{g_k \Delta z}{2} \right) \qquad (5.71)$$

For a positive value of g (i.e. the optical gain) the above expression must be greater than unity. This requires Z_k to be negative, confirming our choice of negative resistance for the model. If our choice of Δz is small enough (i.e. the number of sections M is sufficiently large) the exponential term in equation (5.71) can be expanded by a series. Then it can be shown that for small Δz we have

$$\frac{Z_k}{Z_0} \approx -\frac{g_k \Delta z}{2} \qquad (5.72)$$

A similar equivalent circuit exists for the fields travelling in the $-z$-direction (i.e. v_k), but with the positions of the transmission line and negative resistance reversed (see Figure 5.13). The entire amplifier will therefore be modelled by a cascade of M such equivalent circuits. The exact value of Z_k for each section has to be determined from equation (5.72) using the TMM analysis in which g_{mk} and g_k are determined. The only remaining part of the amplifier to be modelled is the facet. This is in fact a refractive index step and can be modelled

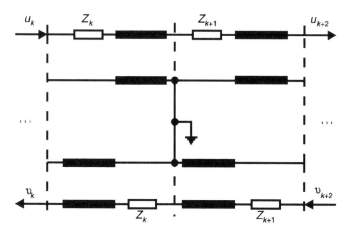

Figure 5.14   Complete equivalent circuit model for a semiconductor laser amplifier

easily by two terminating loads R_{L1} and R_{L2} to account for the reflectivities r_1 and r_2, that is [10, 45]

$$r_1 = \frac{R_{L1} - Z_0}{R_{L1} + Z_0} \tag{5.73}$$

$$r_2 = \frac{R_{L2} - Z_0}{R_{L2} + Z_0} \tag{5.74}$$

The complete equivalent circuit for a SLA is shown in Figure 5.14. This is for a single-section amplifier. To model multi-section SLAs, we first derive an equivalent circuit for each section and then cascade them in a manner similar to cascading the one-dimensional waveguides in the TMM analysis (Figure 5.8). Notice that in theory, when M such sub-networks are cascaded there will be reflections of voltages from later stages due to a mismatch between the characteristic impedance of the transmission line of the kth section and the negative resistance of the $(k + 1)$th section. This is equivalent to the situation which we discussed during the derivation of the TMM, namely that there should be reflections between the interfaces of each section because the optical gain is different in each section, and in theory the refractive indices (the optical gain contributes to the imaginary part; see Chapter 4) are therefore different between each section and reflection coefficients, which are complex, will occur. However, as discussed by Bjork and Nilsson [5], the imaginary parts of the refractive indices are very small (see the discussion on the derivation of the modal gain coefficients with the perturbation method in Chapter 4) and such reflections can be ignored in the TMM calculation. Therefore in the present cascaded equivalent circuit of a SLA, which is derived from the TMM analysis, because of similar mismatches between Z_0 and Z_k the reflections will also be ignored. In practice, this can be achieved by connecting stubs at the interface between the kth and $(k + 1)$th sub-networks, such that matching between each stage occurs and no reflections are observed except at the terminating loads which model the end facets [46].

There are a number of advantages of this equivalent circuit model. First, the model can be treated as an electrical circuit like in many microwave engineering problems. This allows us to analyse more complicated structures for SLAs. Secondly, this equivalent circuit actually allows us to include spontaneous emissions into the analysis, which have been neglected so far in the TMM analysis. This second advantage will be explored in depth in Chapter 6.

5.6 APPLICATIONS

In this section the techniques developed in this chapter for analysis of SLAs will be used to study two particular aspects: (i) the effects of structural parameters on the amplifier gain and (ii) some basic considerations in designing optical communication systems with SLAs as in-line repeaters and pre-amplifiers to optical receivers.

5.6.1 Structural effects on amplifier gain

A SLA with the parameters listed in Table 5.2 has been analysed to study the effects on the amplifier gain of varying the different structural parameters of the amplifier. The pumping ratio i/t_{th} has been fixed at 0.9, where the coupling of spontaneous emissions to the signal mode is unimportant [23]. In this case the TMM can be used to analyse the SLA by taking into consideration the non-uniform photon density distribution and carrier density distribution. The results are shown in Figures 5.15–5.19.

In Figure 5.15 the effect of amplifier length L has been studied. The output power levels of the amplifier are fixed at five different values. It can be seen that for low output power the amplifier gain G can be maximised by adjusting the value of L such that the cavity resonances can match with that of the signal wavelength. As the output power increases, the ripples observed in the curves for lower output power reduce and the value of G falls gradually with increasing L. This is because the amplifier starts to saturate for these values of output power. For $P = 0 \, \text{dBm}$, no ripples occur along the curve and the gain falls rapidly with increasing L as the amplifier is heavily saturated.

The effect of the width of the active layer W on the amplifier gain is shown in Figure 5.16. For low output power levels, changes in W do not affect the amplifier gain significantly, although as W increases we expect that the value of the threshold current simultaneously increases. As the output power increases, it can be seen that narrowing the active region will actually reduce the gain G. The reduction becomes more prominent as W is reduced beyond about $2 \, \mu\text{m}$. Again for $P = 0 \, \text{dBm}$, where the amplifier should be saturated, the reduction in G with reducing W is very clear. In general, it can be concluded that W does not affect the unsaturated gain of a SLA significantly. This allows us to optimise the width of the active region with respect to the following three constraints: (i) the threshold current value for the SLA structure (which

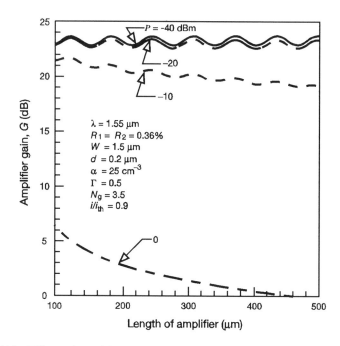

Figure 5.15   Effects of amplifier length on the gain for a SLA analysed by the TMM

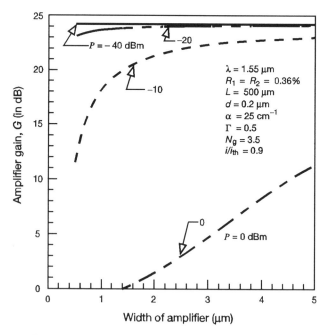

Figure 5.16   Effect of the active layer width on the amplifier gain for a SLA analysed using the TMM

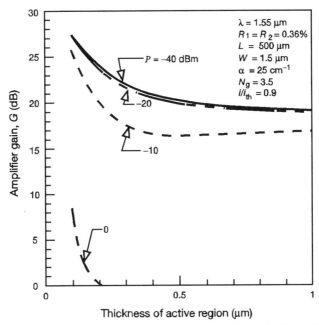

Figure 5.17 Effect of the active thickness on the amplifier gain for a SLA analysed using the TMM

favours a large W), (ii) the saturation output power (which also favours a large W) [63] and (iii) the coupling efficiency [64].

In Figure 5.17 the effect of the active layer thickness d on the amplifier gain G is shown. In the calculations, the change in optical confinement factor Γ with thickness d must be taken into account. We have used the values of Γ calculated for the TE modes in Figure 4.11 of Chapter 4 in the computation, assuming that the structure is under STM operation for all values of d. It can be seen that for all output power levels, the value of G rapidly reduces with d until around $0.35\,\mu m$, beyond which G stays fairly constant. Since a polarisation insensitive SLA will require a thicker active layer region, we can design the SLA in the region of $d > 0.3\,\mu m$, the amplifier gain of which is fairly constant [24]. Ultimately, the upper limit on d is imposed by three constraints: (i) the value of the threshold (which favours a thinner active region), (ii) the thermal resistance of the active region (which also favours a thinner active region) [63] and (iii) the coupling efficiency (which favours a thicker active region to match W in creating a more circular spot size to match that of the single-mode fibre).

Finally, the effects of facet reflectivities on the amplifier gain G are analysed and are shown in Figures 5.18 and 5.19. In Figure 5.19 the SLA under analysis is symmetrical (i.e. $R_1 = R_2$), whereas the results shown in Figure 5.18 are for an asymmetric SLA. In general, it can be seen from these curves that G reduces with increasing R_1 and R_2. This is because the signal wavelength under analysis is not matched with that of the cavity resonance. Otherwise an enhancement of G should be observed. Nevertheless, the amplifier becomes more and more

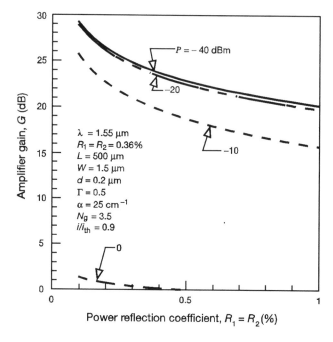

Figure 5.18   Effect of the facet reflection coefficient on the amplifier gain for a SLA analysed using the TMM

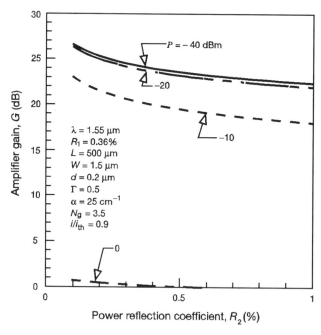

Figure 5.19   As Figure 5.18 but the SLA has an asymmetric mirror facet configuration

sensitive to a mismatch in frequencies, and will saturate much earlier than SLAs with lower reflectivities. Hence, it can be seen that reducing R_1 and R_2 by anti-reflection (AR) coatings is actually more beneficial than trying to match the cavity resonance with the signal wavelength, as the saturation output power can also be improved by low reflectivities.

5.6.2 System considerations

In practical optical fibre communication systems SLAs are usually employed as in-line, non-regenerative repeaters and as pre-amplifiers to optical receivers [39]. In both applications, it is often necessary to design a system such that the optical power incident into the optical receiver will always be maintained above a certain minimum power level P_{min}, such that a minimum signal-to-noise ratio can be maintained below a maximum bit error rate (BER) of B_e [65]. Consider such an optical communication system with M identical SLAs as in-line repeaters, each separated by a distance L_s (see Figure 2.3 in Chapter 2). There is also an identical SLA used as a pre-amplifier to the optical receiver. Each SLA is biased at an identical current i and is maintained at a constant net gain G (i.e. the fibre-to-fibre gain). This can be achieved by controlling the temperature of the SLAs with feedback control by monitoring the output power from each SLA [66]. Assume the system employs intensity modulation (IM) to transmit the data. The SLAs will provide a net gain G at each stage to counteract the losses accumulated in the fibre of length L_s. If the loss of the fiber is α nep/km, then the input power at the output of the first amplifier is $[\eta_1 P_{in} \exp(-\alpha L_s)]G$, where P_{in} is the input optical power and η_1 is the coupling efficiency between the transmitter and the first fibre. Then after M stages, the output optical power from the last section of fibre will be

$$P = \eta_1 P_{in}(G \exp(-\alpha L))^M \tag{5.75}$$

Thus the optical power incident onto the photodetector P_D is

$$P_D = \eta_2 P \tag{5.76}$$

where η_2 is the coupling efficiency between the pre-amplifier and the receiver. There are two major constraints on P_D in practical design. As we have discussed above, $P_D \geqslant P_{min}$ in order to maintain a minimum bit error rate in the receiver. On the other hand, P_D must be smaller than the saturation output power P_{sat} of the pre-amplifier to the receiver to avoid any saturation. Otherwise, G will become saturated and the overall gain is smaller than that depicted in equation (5.75). Such saturation will reduce P_D from the value of P_{min}, and the overall performance of the optical receiver will deteriorate.

In Figure 5.20 we have calculated the value of P_D against M for $L_s = 1$ km, 5 km and 10 km, respectively, with $\eta_1 = \eta_2 = \eta_c = -10$ dB. We have assumed that the unsaturated gain G is maintained at 20 dB and 25 dB by controlling the bias current. In some circumstances the number of SLAs which can be used is also limited (because of, for example, cost). A third constraint which can then be added is $M \leqslant M_0$, as illustrated in Figure 5.20. The bounded region

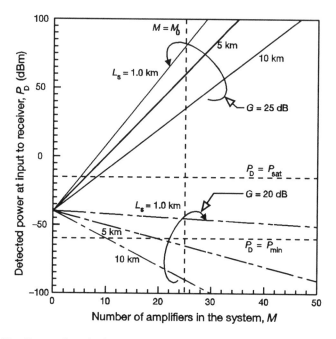

Figure 5.20 Detected optical power at the receiver for different number of repeaters used in the system with different repeater spacings for $G = 20\,\text{dB}$ and $25\,\text{dB}$, respectively

indicates the possible number of amplifiers M used for different L_s at a particular value of P_{in}, which satisfies the constraints of $P_{sat} \geqslant P_D \geqslant P_{min}$ and $M \leqslant M_0$ (see Figure 5.19). When $G = 20\,\text{dB}$, it can be seen that the overall transmission span is lossy and no power gain is achieved at the input to the receiver. However, it thus increases the transmission span from that when $G = 0\,\text{dB}$. It should be noted also that when the system is under such circumstances, the total transmission span ML_s remains constant for all combinations of M and L_s. This is not true when $G = 25\,\text{dB}$, in which there is a net power gain at the input to the optical receiver. In this case the system designer must be careful not to saturate the amplifier chain [67]. Furthermore, the total transmission span for this case is less than that in the former case when $G = 20\,\text{dB}$. Thus it appears to be desirable to operate the present system with a lower gain value. It should be mentioned, however, that there are circumstances in which it may be desirable to operate the amplifiers under saturation. This will be the case when the output power at each stage has to be maintained at a constant level. An examination of the gain G against the input power P_{in} curves calculated in the previous section reveals that when a SLA is operated under saturation conditions, any sudden fall in P_{in} will actually increase the value of the gain value G and hence increase the output power P_{out}. If the design is appropriate, then a constant value of output power P_{out} can be determined in such systems.

It is apparent that the considerations that have to be taken into account in designing systems with SLAs are more complicated than may be expected by common sense, e.g. we may expect a high value of G will be beneficial to the system. In fact, examination of Figure 5.19 indicates that different aspects of the performance of both the system and the amplifiers must be examined thoroughly and carefully.

5.7   ANALYSIS OF GAIN SATURATION IN A SLA WITH A UNIFORM MATERIAL GAIN PROFILE

In subsection 5.2.2 an analytical formula for the gain saturation in a SLA was introduced. In this section a detailed analysis of the formulae will be outlined [26, 30]. We start our formulation with the rate equations, which are described in subsections 5.3.1 and 6.2.2 of Chapters 5 and 6, respectively [13]

$$\frac{\mathrm{d}S_m}{\mathrm{d}t} = -\frac{S_m}{\tau_\mathrm{p}} + \Gamma G_m S_m + \Gamma E_{\mathrm{cf},m} + S_{\mathrm{in},m} \tag{5.77}$$

$$\frac{\mathrm{d}n}{\mathrm{d}t} = P - R - \sum_m G_m \Gamma S_m \tag{5.78}$$

The definitions of the symbols have already been described in subsection 5.3.1. The gain coefficient, g, for the number of photons can be considered as composed of two parts: an unsaturated part g_0, which is independent of photon density, and a saturated part, which depends on the photon density S_m. Mathematically, this can be described by

$$g = g_0 - \frac{\mathrm{d}\Delta g(S_m)}{\mathrm{d}S_m} \cdot S_m \tag{5.79}$$

Inspection of equation (5.78) reveals that the solution of the carrier density will also consist of both an unsaturated part, which is independent of the photon density, and another part, which will depend on the photon density. We can attribute the unsaturated gain to the unsaturated photon density part of the carrier density. At a steady state, the unsaturated part of n will be given by $P\tau_n$ if the carrier recombination rate is given by $R = n/\tau_n$ [69]. Then it can be seen that the gain in the photon number due to the amplifier is given by

$$g_0 = \frac{A\Gamma}{2}(P\tau_n - n_0) \tag{5.80}$$

which is measured in s^{-1}. If the amplifier is pumped towards the threshold, the corresponding value of the threshold gain in photon number for lasing is then given by

$$g_{0\mathrm{th}} = \frac{A\Gamma}{2}(P_{\mathrm{th}}\tau_n - n_0) = \frac{1}{2\tau_\mathrm{p}} \tag{5.81}$$

where the subscript "th" represents quantities at the threshold and τ_p is the lifetime of the photons travelling half of the round trip in the cavity, which are lost via the mirror facets [7]. Combining equations (5.80) and (5.81) results in

$$g_0 = \frac{1}{2\tau_p} [(1 + A\Gamma\tau_p n_0)(P/P_{th} - 1) + 1] = \frac{n_{sp} P_R + 1}{2\tau_p} \quad (5.82)$$

where $P_R = (P/P_{th} + 1)$ is the relative pumping rate supplied to the laser, and $n_{sp} = (1 + A\Gamma\tau_p n_0)$ is the population inversion parameter. If the radiative recombination due to spontaneous emissions is the dominating recombination mechanism, then the photon density can be expressed in terms of pumping rates below and at the lasing threshold by [30]

$$S_m = (P - P_{th})\tau_p = \frac{P_R N_g n_{sp}}{Ac\tau_{sp}} \quad (5.83)$$

If K is the total rate of loss of photon numbers in the laser cavity, then by using equation (5.79) the following inequality can be obtained if the optical amplification is to be maintained:

$$\frac{n_{sp} P_R + 1}{2\tau_p} - \frac{d\Delta g}{dS_m} \cdot \frac{P_R N_g n_{sp}}{Ac\tau_{sp}} \leqslant K \quad (5.84)$$

The amplifier gain saturates when it starts to lase, which occurs when the equality of the above equation holds. In this case K is equal to the threshold gain value given in equation (5.81). Substituting this into the above equation and rearranging gives the saturation parameter [30]

$$\frac{d\Delta g}{dS_m} = \frac{Ac\tau_{sp}}{2N_g V_0 \tau_p} \quad (5.85)$$

If we substitute equations (5.85), (5.83) and (5.82) into equation (5.79), it can be seen that the gain in photon number is halved when the intensity of the optical beam in the amplifier becomes I_{sat}, which is given by [26]

$$I_{sat} = \frac{hf}{A\tau_{sp}} \quad (5.86)$$

When the gain in photon number is halved, it follows that the gain in the power of the optical field is also halved (i.e. drops by 3 dB) as the optical power is proportional to hf. The above expression has an interesting implication on the equivalent circuit representation of spontaneous emissions derived in Chapter 6. It can be shown that the spontaneous emission factor β can be derived from equations (5.77) and (5.78) as [30]

$$\beta = \frac{A\Gamma c\tau_{sp}}{V_e N_g} = \frac{Ac\tau_{sp}}{V_0 N_g} \quad (5.87)$$

If we substitute the expressions for the material gain coefficient $g_m = A(n - n_0)$ and the population inversion parameter $n_{sp} = n/(n - n_0)$ into the single-sided power spectral density of the quantum mechanical fluctuations which are used to model spontaneous emissions in Chapter 6, we can obtain the following expression using equation (5.87):

$$S_e = hf \Gamma g_m n_{sp} \Delta z = hf \Gamma An \Delta z = hf \beta R_{sp}(N_g/c) V_0 \Delta z \quad (5.88)$$

where $R_{sp} = n/\tau_{sp}$. If we measure the power of the spontaneous emission coupled to the lasing mode, it will have a bandwidth $\Delta\omega$ equal to the spectral linewidth below the threshold of the laser [61]. It can be seen from the above equation that by proper adjustment of the length of each cavity Δz (i.e. the Q-factor of each individual section in the amplifier cavity), the relationship between the photon statistical interpretation of the rate equation in terms of β and R_{sp} with the spectral behaviour at the output of the amplifier (which can be measured readily) can be established correctly without the necessity of refining the definitions of terms in the rate equations (as in the critical analysis performed by Sommers [29]).

5.8 SUMMARY

In this chapter an analysis of the signal gain of a semiconductor laser amplifier and the corresponding saturation output power have been described. Simple and basic analysis by assuming a uniform material gain profile along the amplifier has been outlined. Two different approaches to this type of analysis have been compared. The active FP formulation was found to be more suitable than the photon statistics formulation in analysing SLAs, because the phase information of the signals will be lost in the latter approach. To analyse the characteristics of SLAs in a more systematic manner, two different approaches have been reviewed. The rate equations formulation is the basic method as it describes mathematically the interaction between photons and carriers in the amplifier. However, to take into account the actual non-uniform material gain profile in the amplifier, travelling-wave equations have to be added to the basic rate equations, the solutions of which will require iterations and a check for self-consistency at every point along z in the amplifier. The average photon density (AVPD) approximation introduced by Adams *et al.* [37] has been described which, by using the average photon density in the rate equations after solving the travelling-wave equations alone, reduces the number of equations to be solved to three instead of three at every point along z. The limitation of this approximation depends on the accuracy of using the average photon density instead of the actual photon density. In travelling-wave amplifiers, which have very low facet reflectivities, this approximation may not be very accurate. Moreover, the radiative carrier recombination rate due to spontaneous emissions R_{sp} has to be assumed to be independent of z. In practice, since the material gain g_m profile is not z-independent, the actual carrier density n will also be z-dependent and hence R_{sp} should also be spatially dependent. To overcome these shortcomings an alternative solution technique using transfer matrices has been described. This allows R_{sp}, n and g_m to be z-dependent, and the field profile along the amplifier can also be solved. However, it should be noted that this technique only works for index guided SLAs for which effective indices and transverse modal fields can be calculated. Also, the carrier density n is assumed to be independent of x and y as with all the other methods described above. Despite such assumptions, the technique gives an accurate result with a relatively simple algorithm which will always converge. It has been compared with the results calculated by AVPD approximation, and reasonable agreement has been obtained for moderate values of injection current. In this chapter the effect of spontaneous emission on the signal has been ignored. This will be discussed in Chapter 6.

Using the transfer matrix method to analyse SLAs also allows us to derive an equivalent circuit model for them. In this model the active region of the amplifier is modelled as a cascaded network of negative resistance and lossless transmission line. This model will be extremely useful for the analysis and discussion in the next chapter. Finally, the analysis of SLAs has been applied to analyse the structural effects on the gain characteristics of SLAs. These analyses can be used with system design constraints to derive an optimum structure for system applications. The saturation output power was shown to be extremely important in deciding the maximum number of repeaters and the spacing between them in optical communication systems. Proper use of SLAs in optical communication systems can increase the optical power incident onto an optical receiver and hence improve its sensitivity. However, just considering the signal power is not sufficient as the performance of the receiver is determined ultimately by the signal-to-noise ratio at the input of the receiver [66, 68]. In the next chapter the noise characteristics of SLAs will be considered, so that a more complete picture can be drawn.

5.9 REFERENCES

[1]   G. H. B. Thompson, *Physics of Semiconductor Laser Devices*, John Wiley and Sons, New York, 1980.
[2]   J. Buus and R. Plastow, "A theoretical and experimental investigation of Fabry–Perot semiconductor laser amplifiers", *IEEE J. Quantum Electron.*, **QE-21**, No. 6, pp. 614–618, 1985.
[3]   D. Marcuse, "Computer model of an injection laser amplifier", *IEEE J. Quantum Electron.*, **QE-19**, No. 1, pp. 63–73, 1983.
[4]   C. Y. J. Chu and H. Ghafouri-Shiraz, "Equivalent circuit theory of spontaneous emission power in semiconductor laser optical amplifiers", *IEEE, J. Lightwave Technol.*, **LT-12**, No. 5, pp. 760–767, 1994.
[5]   G. Bjork and O. Nilsson, "A new exact and efficient numerical matrix theory of complicated laser structures: properties of asymmetric phase-shifted DFB lasers", *J. Lightwave Technol.*, **LT-5**, No. 1, pp. 140–146, 1987.
[6]   C. Jørgensen, N. Storkfelt, T. Durhuus, B. Mikkelsen, K. E. Stubkjaer, B. Fernier, G. Gelly and P. Doussiere, "Two-section semiconductor optical amplifier used as an efficient channel dropping node", *IEEE Photon. Technol. Lett.*, **4**, No. 4, pp. 348–350, 1992.
[7]   A. Yariv, *Optical Electronics*, 3rd edition, Holt–Saunders, 1985.
[8]   M. J. Adams, *An Introduction to Optical Waveguides*, John Wiley and Sons, New York, 1981.
[9]   H. A. Haus, *Wave and Fields in Optoelectronics*, Prentice-Hall, New York, 1984.
[10] C. A. Balanis, *Advanced Engineering Electromagnetics*, John Wiley and Sons, New York, 1989.
[11] Y. Yamamoto, "Characteristics of AlGaAs Fabry–Perot cavity type laser amplifiers", *IEEE J. Quantum Electron.*, **QE-16**, No. 10, pp. 1047–1052, 1980.
[12] T. Mukai and Y. Yamamoto, "Gain, frequency bandwidth, and saturation output power of AlGaAs DH laser amplifiers", *IEEE J. Quantum Electron.*, **QE-17**, No. 6, pp. 1028–1034, 1981.

[13] Y. Mukai, Y. Yamamoto and T. Kimura, "Optical amplification by semi-conductor lasers", *Semiconductors and Semimetals*, **22**, Part E, pp. 265–319, Academic Press, London, 1985.

[14] J. Wang, H. Olesen and K. E. Stubkjaer, "Recombination, gain and bandwidth characteristics of $1.3\,\mu m$ semiconductor laser amplifiers", *IEEE J. Lightwave Technol.*, **LT-5**, No. 1, pp. 184–189, 1987.

[15] T. Saitoh, T. Mukai and O. Mikami, "Theoretical analysis and fabrication of antireflection coatings on laser-diode facets", *J. Lightwave Technol.*, **LT-3**, No. 2, pp. 288–293, 1985.

[16] T. Saitoh and T. Mukai, "Recent progress in semiconductor laser amplifiers", *IEEE J. Lightwave Technol.*, **LT-6**, No. 11, pp. 1156–1164, 1988.

[17] T. Saitoh and M. Mukai, "$1.5\,\mu m$ GaInAs travelling-wave semiconductor laser amplifier", *IEEE J. Quantum Electron.*, **QE-23**, No. 6, pp. 1010–1020, 1987.

[18] G. P. Agrawal and N. K. Dutta, *Long-wavelength Semiconductor Lasers*, Van-Nostrand Reinhold, New York, 1986.

[19] T. Mukai and Y. Yamamoto, "Noise in an AlGaAs semiconductor laser amplifier", *IEEE J. Quantum Electron.*, **QE-18**, No. 4, pp. 564–575, 1982.

[20] T. Mukai, Y. Yamamoto and T. Kimura, "S/N and error rate performance in AlGaAs semiconductor laser preamplifier and linear repeater systems", *IEEE Trans. Microwave Theory Technol.*, **MTT-30**, No. 10, pp. 1548–1556, 1982.

[21] R. Loudon, *The Quantum Theory of Light*, 2nd edition, Oxford University Press, Oxford, UK, 1983.

[22] D. T. Cassidy, "Comparison of rate-equation and Fabry–Perot approaches to modelling a diode laser", *Appl. Opt.*, **22**, No. 21, pp. 3321–3326, 1983.

[23] M. J. Adams, "Time dependent analysis of active and passive optical bistability in semiconductors", *IEE Proc.*, **132**, Part J, No. 6, pp. 343–348, 1985.

[24] T. Saitoh and T. Mukai, "Structural design for polarization-insensitive travelling-wave semiconductor laser amplifiers", *Opt. Quantum Electron.*, **21**, pp. S47–S48, 1989.

[25] P. J. Stevens and T. Mukai, "Predicted performance of quantum-well GaAs-(GaAl)As optical amplifiers", *IEEE J. Quantum Electron.*, **26**, No. 11, pp. 1910–1917, 1990.

[26] Y. Yamamoto, S. Saito and T. Mukai, "AM and FM quantum noise in semiconductor lasers — Part II: Comparison of theoretical and experimental results for AlGaAs lasers", *IEEE J. Quantum Electron.*, **QE-19**, No. 1, pp. 47–58, 1983.

[27] D. Marcuse, "Classical derivation of the laser rate equation", *IEEE J. Quantum Electron.*, **QE-19**, No. 8, pp. 1228–1231, 1983.

[28] I. D. Henning, M. J. Adams and J. V. Collins, "Performance prediction from a new optical amplifier model", *IEEE J. Quantum Electron.*, **QE-21**, No. 6, pp. 609–613.

[29] H. S. Sommers Jr., "Critical analysis and correction of the rate equation for the injection laser", *IEE Proc.*, **132**, Part J, No. 1, pp. 38–41, 1985.

[30] Y. Yamamoto, "AM and FM quantum noise in semiconductor lasers — Part I: Theoretical analysis", *IEEE J. Quantum Electron.*, **QE-19**, No. 1, pp. 34–46, 1983.

[31] T. Mukai, K. Inoue and T. Saitoh, "Homogeneous gain saturation in $1.5\,\mu m$ InGaAsP travelling-wave semiconductor laser amplifiers", *Appl. Phys. Lett.*, **51**, No. 6, pp. 381–383, 1987.

[32] J. Buus, "Principles of semiconductor laser modelling", *IEE Proc.*, **132**, Part J, No. 1, pp. 42–51, 1985.

[33] H. Ghafouri-Shiraz and C. Y. J. Chu, "Effect of phase shift position on spectral linewidth of the $\pi/2$ distributed feedback laser diode", *J. Lightwave Technol.*, **LT-8**, No. 7, pp. 1033–1038, 1990.

[34] M. I. Sargent, M. O Scully and W. E. Lamb Jr, *Laser Physics*, Addison-Wesley, Wokingham, UK, 1974.

[35] A. J. Lowery, "Modelling spectral effects of dynamic saturation in semiconductor laser amplifiers using the transmission-line laser model", *IEE Proc.*, **136**, Part J, No. 6, pp. 320–324, 1989.

[36] J. Buus, "The effective index method and its application to semiconductor lasers", *IEEE J. Quantum Electron.*, **QE-18**, No. 7, pp. 1083–1089, 1982.

[37] M. J. Adams, J. V. Collins and I. D. Henning, "Analysis of semiconductor laser optical amplifiers", *IEE Proc.*, **132**, Part J, pp. 58–63, 1985.

[38] T. Makino and T. Glinski, "Transfer matrix analysis of the amplified spontaneous emission of DFB semiconductor laser amplifiers", *IEEE J. Quantum Electron.*, **QE-24**, No. 8, pp. 1507–1518, 1988.

[39] G. Eisenstein, "Semiconductor optical amplifiers", *IEEE Circuits and Devices Mag.*, pp. 25–30, 1989.

[40] S. Banerjee and A. Sharma, "Propagation characteristics of optical waveguiding structures by direct solution of the Helmholtz equation for total fields", *J. Opt. Soc. Am., A*, **6**, No. 12, pp. 1884–1894, 1989.

[41] C. H. Henry and B. H. Verbeck, "Solution of the scalar wave equation for arbitrarily shaped dielectric waveguides by two-dimensional Fourier analysis", *IEEE J. Lightwave Technol.*, **LT-7**, No. 32 pp. 308–313, 1989.

[42] P. B. Johns and R. L. Beurle, "Numerical solution of 2-dimensional scattering problems using a transmission-line matrix", *IEE Proc.*, **118**, No. 9, pp. 1203–1208, 1971.

[43] W. J. R. Hoefer, "The transmission-line matrix method-theory and applications", *IEEE Trans. Microwave Theory Technol.*, **MTT-33**, No. 10, pp. 882–893, 1985.

[44] W. J. R. Hoefer, "The discrete time domain Green's function or Johns matrix-anew powerful concept in transmission line modelling (TLM)", *Int. J. Numer. Model*, **2**, pp. 215–225, 1989.

[45] A. J. Lowery, "Transmission-line modelling of semiconductor lasers: the transmission-line laser model", *Int. J. Numer. Model*, **2**, pp. 249–265, 1989.

[46] A. J. Lowery, "New in-line wideband dynamic semiconductor laser amplifier model", *IEE Proc.*, **135**, Part J, No. 3, pp. 242–250, 1988.

[47] N. Wax, *Selected Papers on Noise and Stochastic Process*, Dover, New York, 1954.

[48] A. J. Lowery, "Amplified spontaneous emission in semiconductor laser amplifiers: validity of the transmission-line laser model", *IEE Proc.*, **137**, Part J, No. 4, pp. 241–247, 1990.

[49] G. Bjork and O. Nilsson, "A tool to calculate the linewidth of complicated semiconductor lasers", *IEEE J. Quantum Electron.*, **QE-23**, No. 8, pp. 1303–1313, 1987.

[50] I. Maio, "Gain saturation in travelling-wave ridge waveguide semiconductor laser amplifiers", *IEEE Photon. Technol. Lett.*, **3**, No. 7, pp. 629–631, 1991.

[51] P. C. Kendall, M. J. Adams, S. Ritichie and M. J. Robertson, "Theory for calculating approximate values for the propagation constants of an optical rib waveguide by weighting the refractive indices", *IEE Proc.* **134**, Part A, No. 8, pp. 699–702, 1987.

[52] M. Cross and M. J. Adams, "Waveguiding properties of stripe-geometry double heterostructure injection lasers", *Solid State Electron.*, **15**, pp. 919–921, 1972.

[53] H. A. Haus and S. Kawakami, "On the excess spontaneous emission factor in gain-guided laser amplifiers", *IEEE J. Quantum Electron.*, **QE-21**, No. 1, pp. 63–69, 1985.

[54] R. M. Fortenberry, A. J. Lowery and R. S. Tucker, "Up to 16 dB improvement in detected voltage using two section semiconductor optical amplifier detector", *Electron. Lett.*, **28**, No. 5, pp. 474–476, 1992.

[55] P. M. Morse and H. Feshbach, *Methods of Theoretical Physics: Part I*, McGraw-Hill, New York, 1953.

[56] S. G. Mikhlin and K. L. Smolitskiy, *Approximate Methods for Solution of Differential and Integral Equations*, American Elsevier Publishing Co., 1967.

[57] C. Y. J. Chu and H. Ghafouri-Shiraz, "Structural effects on polarization sensitivity of travelling wave semiconductor laser amplifiers", *3rd Bangor Communications Symposium*, UK, pp. 19–22, 1991.

[58] C. D. Zaglanakis and A. J. Seeds, "Computer model for semiconductor laser amplifiers with RF intensity-modulated inputs", *IEE Proc.*, **139**, Part J, No. 4, pp. 254–262, 1992.

[59] R. S. Tucker, "Circuit model of double-heterostructure laser below threshold", *IEE Proc.*, **128**, Part I, No. 3, pp. 101–106, 1981.

[60] O. Nilsson, Y. Yamamoto and S. Machida, "Internal and external field fluctuations of a laser oscillator: Part II — Electrical circuit theory", *IEEE J. Quantum Electron.*, **QE-22**, No. 10, pp. 2043–2051, 1986.

[61] C. H. Henry, "Theory of the phase noise and power spectrum of a single mode injection laser", *IEEE J. Quantum Electron.*, **QE-19**, No. 9, pp. 1391–1397, 1983.

[62] C. H. Henry, "Theory of spontaneous emission noise in open resonators and its application to lasers and optical amplifiers", *IEEE J. Lightwave Technol.*, **LT-4**, No. 3, pp. 288–297, 1986.

[63] J. C. Simon, "GaInAsP semiconductor laser amplifiers for single-mode fiber communications", *J. Lightwave Technol.*, **LT-5**, No. 9, pp. 1286–1295, 1987.

[64] J. John, T. S. M. Maclean, H. Ghafouri-Shiraz and J. Niblett, "Matching of single-mode fibre to laser diode by microlenses at $1.5\,\mu m$ wavelength", *IEE Proc.*, **141**, Part J, No. 3, pp. 178–184, 1994.

[65] T. Li, *Topics in Lightwave Transmission Systems*, Academic Press, London, 1991.

[66] H. Kressel, "Semiconductor devices for optical communications", 2nd edition, Springer-Verlag, Berlin, 1982.

[67] A. Elrefaire and C. Lin, "Performance degradations of multi gigabit-per-second NRZ/RZ lightwave system due to gain saturation in travelling-wave semiconductor optical amplifiers", *IEEE Photon. Technol. Lett.*, **1**, No. 10, pp. 300–303, 1989.

[68] H. Taub and D. L. Schilling, *Principles of Communication Systems*, 2nd edition, McGraw-Hill, New York, 1986.

[69] R. Olshansky, C. B. Su, J. Manning and W. Powazink, "Measurement of radiative and nonradiative recombination rates in InGaAsP and AlGaAs light sources", *IEEE J. Quantum Electron.*, **QE-20**, No. 8, pp. 838–854, 1984.

[70] C. Y. J. Chu and H. Ghafouri-Shiraz, "Analysis of gain and saturation characteristics of a semiconductor laser optical amplifier using transfer matrices", *IEEE J. Lightwave Technol.*, **12**, No. 8, pp. 1378–1386, 1994.

[71] C. Y. J. Chu and H. Ghafouri-Shiraz, "A simple method to determine carrier recombinations in a semiconductor laser optical amplifier", *IEEE Photon. Technol. Lett.*, **5**, No. 10, pp. 1182–1185, 1993.

ANALYSIS AND MODELLING OF SEMICONDUCTOR LASER AMPLIFIERS: NOISE CHARACTERISTICS

6.1 INTRODUCTION

In Chapter 5 we discussed the analysis of the gain and saturation characteristics of semiconductor laser amplifiers (SLAs). The fairly complex interaction between injected carriers and photons (which include both signal and amplified spontaneous emissions) can be described by a set of rate equations and a travelling-wave equation. As discussed in Chapter 2, spontaneous emission is a random process which is statistically stationary [1, 2] and its ensemble average (i.e. power) can be measured [3]. In fact, spontaneous emission in optical devices is like shot noise in electronic circuits [4] and it is a source of noise in SLAs. In electronic amplifiers the noise generated by the various components is measured as the output power in the absence of a signal input. As the noise processes are uncorrelated with the signal input, the noise power is directly added to the power output owing to the signal.

The above analogy between shot noise in electronic amplifiers and spontaneous emissions in SLAs reveals an additional point. That is, spontaneous emission will cause random fluctuations in both amplitude and phase of the signal and, as in electronic communication systems, this may give rise to error in signal detection. Spontaneous emissions can influence the performance of SLAs in three ways [5]:

(i) Because spontaneous emission will involve the transition of carriers, its random nature will induce fluctuations in the carrier density n.

(ii) The fluctuation in n will induce fluctuations in the material gain coefficient g_m and hence affect both the amplitude and the phase of the signal, as well as the total number of photons.

(iii) The fluctuations in n will also induce changes in the refractive index N_g as predicted by the Kramers–Kronig relation, causing further phase changes. This is similar to the physical process behind linewidth broadening in semiconductor lasers [6, 7].

The above three aspects are fluctuations induced by spontaneous emissions via the carrier density n. In addition, the spontaneous emission photons can interact directly with the signal. As we shall see later, this interaction can be enhanced during the detection process, introducing a new type of noise in the detector known as the *beat noise* [8].

It can be seen that the analysis of noise processes in SLAs is fairly complicated owing to the interaction between different material parameters with photons. This chapter is devoted to such an analysis. Some conventional techniques for dealing with such problems are based on statistical methods, which will be described first. We will then utilise the equivalent circuit model developed in Chapter 5 to analyse spontaneous emissions from a SLA. This approach, first reported here, allows us to investigate more complex structural effects on the spontaneous emissions from SLAs, e.g. the effects of stray reflections from the ends of optical fibres. Finally, as in Chapter 5, the analysis developed here will be used to investigate its implications for the design of low-noise SLAs and the performance of optical communication systems.

6.2 FORMULATION OF NOISE IN SEMICONDUCTOR LASER AMPLIFIERS

Although noise is one of the most important properties of SLAs, it is also the most complicated and difficult to analyse. This is partly because of the mathematical complexity of describing the statistical processes which take place inside the amplifier (see Figure 6.1). There are therefore several different approaches with different degrees of complexity in their formulation. Here we will discuss three of them.

6.2.1 Photon statistics formulation

Mukai and Yamamoto [8, 9] described the application of photon statistic master equations to calculate the noise properties of a SLA. This approach considers the statistical properties of the number of photons in a SLA. The statistical *momentum*, defined as the time derivative of the ensemble average of the rth power of the photon number, is evaluated by analogy with the statistics of population changes of different organisms in an ecological system due to birth and death. A simplified formulation, based on first- and second-order

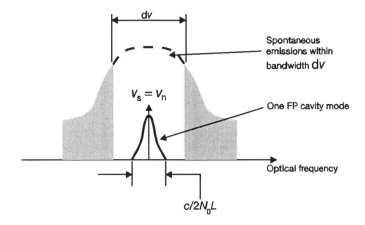

Figure 6.1 Gain spectrum of a FPA and the corresponding spontaneous emission noise spectrum

moments of the photon number N_p in a unit optical bandwidth, has been given by [10]:

$$\frac{d\langle N_p \rangle}{dt} = (A - B - C)\langle N_p \rangle + A \tag{6.1}$$

$$\frac{d\langle N_p^2 \rangle}{dt} = 2(A - B - C)\langle N_p^2 \rangle + (3A + B + C)\langle N_p \rangle + A \tag{6.2}$$

where A is the stimulated emission probability per second, B is the absorption probability per second which both describe the change in photon numbers owing to these processes [11]. C will account for other loss mechanisms (e.g. scattering into claddings, intervalence band absorption [12]), which give rise to a finite photon lifetime τ_p. In equations (6.1) and (6.2), the total number of photons N_p owing to coupling to the output and input have been included in these equations, and hence the only possibility of a change in the photon density is owing to the "birth" of photons owing to radiative transitions and the "death" of photons owing to absorption and scattering (compare this picture with an ecological system in which boundaries are drawn in such a way that there is no net migration into and out of the boundaries, and hence the only possible reasons for a change in the population of organisms in the system will be owing to "birth" and "death" [13]). Mukai and Yamamoto [8] postulated that the coefficients A, B and C are given by (see also subsection 5.2.1 of Chapter 5)

$$A = E_{cv,m} \tag{6.3a}$$

$$B = E_{vc,m} \tag{6.3b}$$

$$C = \frac{1}{\tau_p} \tag{6.3c}$$

for the mth mode. The subscripts "cv" and "vc" represent the direction of radiative transitions, with c representing the conduction band, and v representing the valence band. Strictly speaking, the above equation holds only for the photon number $N_{p,m}$ of the mth mode. In general, the rth order momentum of $N_{p,m}$ is given by [8]

$$
\frac{d\langle N_{p,m}^r \rangle}{dt} =
$$

$$
\sum_{j=0}^{r-1} \binom{r}{j} [\langle (N_{p,m} + 1) N_{p,m}^j \rangle E_{cv,m} + (-1)^{r+j} \langle N_{p,m}^{j+1} \rangle E_{vc,m} + (-1)^{r+j} \langle N_{p,m}^{j+1} \rangle / \tau_p]
$$

$$(6.4)$$

Physically, the photon statistics master equations (6.1) and (6.3) explain how the mean and variance of the photon population of the amplifier are affected by radiative transitions. The interaction between stimulated emissions, spontaneous emissions and absorption will result in a net optical gain as well as fluctuations in the photon number, as indicated by the finite value of the variance $\langle N_{p,m}^2 \rangle$. Notice from equations (6.3) and (6.4) that the population inversion parameter n_{sp} can be defined in terms of the coefficients A and B by [14]

$$
n_{sp} = \frac{A}{A - B} \qquad (6.5)
$$

We can also derive the amplifier gain G in terms of the coefficients A, B and C [10]. Consider the number of photons $N_p(z)$ inside the amplifier. The change in photon population for the *signal* along a distance dz is given by

$$
dN_p = N_g N_p (A - B - C)(dz/c) \qquad (6.6)
$$

where N_g is the group effective refractive index of the medium and c is the velocity of light in free space. In equation (6.6) the quantity $(A - B - C)N_p$ represents the net change in photon number due to stimulated emissions, absorption and other loss mechanisms. The term dz/c represents the time elapsed as the photons travel from z to $z + dz$, i.e. the frame of reference has been transformed from a stationary one with respect to the SLA to one that is moving along the amplifier with the photons at a group velocity of c/N_g. Integrating over a length L and imposing the boundary conditions of the initial number of signal photons, the gain can be derived from equation (6.6) using the definition discussed in subsection 5.2.1 as

$$
G = \frac{N_p(t = LN_g/c)}{N_p(t = 0)} = \exp \left[N_g(A - B - C) \frac{L}{c} \right] \qquad (6.7)
$$

Using equation (6.7) we can solve the first-order differential equation (6.1) to obtain the mean photon number as

$$
\langle N_p \rangle = G\langle N_{p0} \rangle + \frac{(G - 1)A}{A - B - C} \qquad (6.8)
$$

where we have used N_{p0} to represent $N_p(t = 0)$. With the assumption of $C \ll A$ and B, the term $A/(A - B - C)$ is approximately equal to n_{sp} (see equation (6.5)) and hence equation (6.8) can be written as

$$\langle N_p \rangle = G\langle N_{p0} \rangle + (G - 1)n_{sp} \tag{6.9}$$

Using a similar technique, the variance of the photon number at the output of the amplifier can be found by using equations (6.2) and (6.9):

$$\langle N_p^2 \rangle = G\langle N_{p0} \rangle + (G - 1)n_{sp} + 2G(G - 1)n_{sp}\langle N_{p0} \rangle$$
$$+ (G - 1)^2 n_{sp}^2 + G^2[\langle N_{p0}^2 \rangle - \langle N_{p0} \rangle^2 - \langle N_{p0} \rangle] \tag{6.10}$$

In SLAs there can be several possible transverse modes for each longitudinal cavity resonance frequency (see equation (5.1), subsection 5.2.1 in Chapter 5). If there are m_t transverse modes for each longitudinal cavity resonance of bandwidth Δf, then the equivalent noise bandwidth of the amplifier output is given by $m_t \Delta f$ [3, 8, 10]. Then equations (6.8) and (6.10) can be written as

$$\langle N_p \rangle = G\langle N_{p0} \rangle + (G - 1)n_{sp} m_t \Delta f \tag{6.11}$$

$$\langle N_p^2 \rangle = G\langle N_{p0} \rangle + (G - 1)n_{sp} m_t \Delta f + 2G(G - 1)n_{sp}\langle N_{p0} \rangle$$
$$+ (G - 1)^2 n_{sp}^2 m_t \Delta f + G^2[\langle N_{p0}^2 \rangle - \langle N_{p0} \rangle^2 - \langle N_{p0} \rangle] \tag{6.12}$$

In equation (6.11) the first term represents the average photon number owing to the amplified input, whereas the second term represents that owing to spontaneous emissions [8]. The five terms in equation (6.12) represent shot noise owing to the amplified signal, shot noise owing to spontaneous emissions, beat noise between signal and spontaneous emissions, beat noise between spontaneous emission components and excess noise owing to incoherence of the input signal when the output photons are detected by a photodetector, respectively. For a purely coherent input signal (i.e. without phase noise), the last term vanishes [10, 13, 15]. The relative noise power detected by a photodetector (PD) with a quantum efficiency of unity, unit load resistance and unit detection bandwidth can be found from the variance of the output photon number σ_{out}^2 [1, 8–16]:

$$\langle i_n^2 \rangle = e^2 \sigma_{out}^2 = e^2[\langle N_p^2 \rangle - \langle N_p \rangle^2] \tag{6.13}$$

where e is the electron charge. The above formulation using photon statistics is general. We will illustrate briefly how to apply it to the analysis of a simple semiconductor Fabry–Perot amplifier (FPA). Analyses of more complex structures and fibre amplifiers using equations (6.11) and (6.12) follow similar procedures to those outlined below.

Consider the gain spectrum of a FPA with a uniform material gain profile as illustrated in Figure 6.1 [10]. For an optical frequency range of df, the number of spontaneous emission noise photons emitted by this FPA of length L per second is given by (see Section 6.5) [17]

$$\overline{N}_p(f)\mathrm{d}f = \frac{(1 + R_1 G_s)(1 - R_2)(G_s - 1)n_{sp}}{|1 - G_s\sqrt{R_1 R_2}\,\exp(-2j\beta_z L)|^2}\, m_t\,\mathrm{d}f \qquad (6.14)$$

where $\overline{N}_p(f)$ is the single-sided spectral density of photons emitted per second, G_s is the single pass gain of the amplifier, and R_1 and R_2 are the power reflection coefficients of the facets. For one FP cavity resonance with m_t transverse modes, the number of spontaneous emission photons emitted per second is given by

$$\int_{f_n - c/4L}^{f_n + c/4L} \overline{N}_p(f)\mathrm{d}f = \frac{(1 + R_1 G_s^n)(1 - R_2)(G_s^n - 1)n_{sp}\,m_t}{1 - R_1 R_2(G_s^n)^2}\left(\frac{c}{2L}\right) \qquad (6.15)$$

where c is the velocity of light in the amplifying medium, f_n is the nth resonance frequency and G_s^n is the corresponding single pass gain of the amplifier at $f = f_n$ [18]. For a signal of frequency f_s matched to one of the cavity resonances, the gain of the signal is given by equation (5.8) (see subsection 5.2.1 in Chapter 5), that is

$$G = \frac{(1 - R_1)(1 - R_2)G_s^0}{(1 - G_s^0\sqrt{R_1 R_2})^2} \qquad (6.16)$$

where G_s^0 is the single pass gain at $f = f_s$. Then from equation (6.11) it can be shown that by using equations (6.15) and (6.16) we can obtain [10]

$$\langle N_p \rangle = \frac{(1 - R_1)(1 - R_2)G_s^0}{(1 - G_s^0\sqrt{R_1 R_2})}\langle N_{p0}\rangle$$
$$+ \sum_n \frac{(1 + R_1 G_s^n)(1 - R_2)(G_s^n - 1)n_{sp}\,m_t}{1 - R_1 R_2(G_s^n)^2}\left(\frac{c}{2L}\right) \qquad (6.17)$$

where the summation is taken over all longitudinal cavity resonances within Δf, assuming that the output of the SLA is filtered by an optical filter with a pass band bandwidth of Δf (Figure 6.1) and $\langle N_{p0}\rangle$ is the average number of photons injected into the amplifier per second. By substituting equations (6.15), (6.16) and (6.17) into (6.12) and (6.13) and noticing the physical meanings of the terms in equation (6.12), it can be shown that

$$\langle N_p^2 \rangle - \langle N_p\rangle^2 = \langle N_p\rangle + I_1 + I_2$$
$$+ \left[\frac{(1 - R_1)(1 - R_2)G_s^0}{(1 - G_s^0\sqrt{R_1 R_2})^2}\right]^2 (\langle N_{p0}^2\rangle - \langle N_{p0}\rangle^2 - \langle N_{p0}\rangle) \qquad (6.18)$$

where $\langle N_p\rangle$ is given by equation (6.17). In the above derivations we have assumed that the photo-detector has a low pass filter characteristic [19] such that the high frequency beating noises between the signals and those between spontaneous emission components are filtered off for the $\langle N_p\rangle^2$ term [8]. The term I_1 will involve beating between signal and spontaneous emissions [10]. Therefore from equation (6.14), I_1 can be found by the following integral:

$$I_1 = 2G\langle N_{p0}\rangle \int_{f_n-c/4L}^{f_n+c/4L} \overline{N}_p(f)\delta(f-f_s)\,df$$

$$= \frac{2(1-R_1)(1-R_2)^2(1+R_1G_s^0)(G_s^0-1)G_s^0 n_{sp}}{(1-G_s^0\sqrt{R_1R_2})^4} \langle N_{p0}\rangle \qquad (6.19)$$

where $\delta(f-f_s)$ is the Dirac delta function for $f=f_s$. In this case $m_t = 1$ because the input signal is assumed to be in a single state of polarisation and hence the beating between the signal and spontaneous emissions can only take place along that particular transverse mode. The term I_2 can be found by considering the number of photons due to beating between spontaneous emission components within one longitudinal mode, and then summed over Δf:

$$I_2 = \sum_n \int_{f_n-c/4L}^{f_n+c/4L} \overline{N}_p^2(f)\,df \qquad (6.20)$$

Substituting equation (6.14) into the above expression, and after some algebraic manipulation, the integral can be evaluated analytically, which results in

$$I_2 = \sum_n \frac{(1+R_1G_s)^2(1-R_2)^2(G_s-1)^2(1+R_1R_2G_s^2)n_{sp}^2 m_t}{(1-R_1R_2G_s^2)^3} \left(\frac{c}{2L}\right) \qquad (6.21)$$

where we have dropped the superscript n on the single pass gain G_s^2, and hereafter, for simplicity, we assume that the dependence of G_s on the cavity resonance number n is implicit. The relative noise power detected at the output of the FPA is given by equation (6.13). By combining the above results in equations (6.18)–(6.21) we can obtain the total noise power as [10]:

$$\begin{aligned}
\langle i_n^2\rangle/e^2 &= \langle N_p^2\rangle - \langle N_p\rangle^2 \\
&= \frac{(1-R_1)(1-R_2)G_s^0}{(1-G_s^0\sqrt{R_1R_2})^2}\langle N_{p0}\rangle \\
&\quad + \sum_n \frac{(1+R_1G_s)(1-R_2)(G_s-1)n_{sp}m_t}{1-R_1R_2G_s^2}\left(\frac{c}{2L}\right) \\
&\quad + 2\frac{(1-R_1)(1-R_2)^2(1+R_1G_s^0)(G_s^0-1)G_s^0 n_{sp}}{(1-G_s^0\sqrt{R_1R_2})^4}\langle N_{p0}\rangle \\
&\quad + \sum_n \frac{(1+R_1G_s)^2(1-R_2)^2(G_s-1)^2(1+R_1R_2G_s^2)n_{sp}^2 m_t}{(1-R_1R_2G_s^2)^3}\left(\frac{c}{2L}\right) \\
&\quad + \left[\frac{(1-R_1)(1-R_2)G_s^0}{(1-G_s^0\sqrt{R_1R_2})^2}\right]^2 (\langle N_{p0}^2\rangle - \langle N_{p0}\rangle^2 - \langle N_{p0}\rangle)
\end{aligned}$$

$$(6.22)$$

It can be shown that by substituting equations (6.11) and (6.12) into (6.13), the above equation can be written intuitively as

$$\langle i_n^2 \rangle / e^2 = G\langle N_{p0} \rangle + (G-1)n_{sp}m_t \Delta f_1$$
$$+ 2G(G-1)n_{sp}\chi\langle N_{p0} \rangle + (G-1)^2 n_{sp}^2 m_t \Delta f_2$$
$$+ G^2(\langle N_{p0}^2 \rangle - \langle N_{p0} \rangle^2 - \langle N_{p0} \rangle) \tag{6.23}$$

where Δf_1 is the equivalent noise bandwidth of spontaneous emission shot noise and Δf_2 is the equivalent noise bandwidth of the beat noise between spontaneous emission components. Both Δf_1 and Δf_2 can be found by comparing equation (6.22) with (6.23), that is [10]

$$\Delta f_1 = \frac{+\sum_n \dfrac{(1+R_1 G_s)(1-R_2)(G_s-1)}{1-R_1 R_2 G_s^2}\left(\dfrac{c}{2L}\right)}{(G-1)} \tag{6.24}$$

$$\Delta f_2 = \frac{+\sum_n \dfrac{(1+R_1 G_s)^2(1-R_2)^2(G_s-1)^2}{(1-R_1 R_2 G_s^2)^2}\left(\dfrac{c}{2L}\right)}{(G-1)^2} \tag{6.25}$$

It is interesting to see that the physical origins of Δf_1 and Δf_2 are due to the enhancement of noise by the multi-longitudinal cavity modes within the frequency band Δf. The resulting increase in spontaneous emission noise can be accounted for by a proportionate change in bandwidth, with the new net bandwidths given by equations (6.24) and (6.25). The parameter χ in equation (6.23) is known as the *excess noise coefficient* [11, 20] which arises because of the enhancement of signal-spontaneous emissions beating noise by the FP cavity. It can be shown that

$$\chi = \frac{(1-R_1)(1-R_2)(G_s^0-1)}{(1-G_s^0\sqrt{R_1 R_2})^2(G-1)} \tag{6.26}$$

Hence, for a travelling-wave amplifier (TWA) with $R_1 = R_2 = 0$ and $G = G_s^0$ the value of $\chi = 1$. For a FPA with R_1 and $R \neq 0$ the value of χ can be greater than unity, depending on the combinations of the facet reflectivities, material gain and cavity length [11]. It can also be seen that for a TWA, the corresponding noise power can be found simply by substituting $G = G_s^0$ and $\chi = 1$ into equation (6.23), with $\Delta f_1 = \Delta f_2 = \Delta f$. Obviously, the noise emitted from a TWA can be lower than that from a FPA [21]. The noise figure F is defined as the degradation of the signal-to-noise ratio of the SLA at the output (see subsection 2.3.4 in Chapter 2). As we shall see later, when G is large, the signal-spontaneous beating will dominate the overall detected noise power [11]. In this case, it can be shown that for $G \gg 1$

$$F = 2n_{sp}\chi \tag{6.27}$$

For a perfect population inversion with $n_{sp} = 1$ and $\chi = 1$, we arrive at the minimum noise figure of 3 dB, which was derived in an alternative way in subsection 2.5.4 of Chapter 2. The above approach of photon statistic master equations allows us to visualise the physical origins of the noise power detected by the receiver. It also allows us to separate the different contributions to the total noise power by equation (6.12) in the analysis [11]. However, the analysis neglects the spatial dependence of the spontaneous emissions and signal, as well as losing phase information by using the concept of photon counting [22–24]. Nevertheless, its relative simplicity is found to be very useful in analysing noise in SLAs, and subsequent improvements on this technique have been used. A recent application of this approach to noise analysis can be found in the work of Goldstein and Teich [13] who extended the preceding analysis to a general resonator configuration, which included coupling effects between amplifier output and PDs [25].

6.2.2 Rate equation approach

A more direct way to analyse noise in SLAs is to include fluctuations in the multi-mode rate equations used to analyse the semiconductor laser amplifier (see subsection 5.3.1 in Chapter 5). Mukai and Yamamoto [8] used this approach by postulating that the carrier density n and photon number $N_{p,m}$ of the mth mode are related by the following equations:

$$\frac{dn}{dt} = P - R - \sum_m \Gamma G_m N_{p,m} + F_e(t) \tag{6.28}$$

$$\frac{dN_{p,m}}{dt} = -\frac{N_{p,m}}{\tau_p} + \Gamma G_m N_{p,m} + \Gamma E_{cv,m} + P_{in,m} + F_{p,m}(t) \tag{6.29}$$

where P is the pumping rate due to the injection electrical current, R is the carrier recombination rate, G_m is the stimulated emission coefficient given by $E_{cv,m} - E_{vc,m}$, with $E_{cv,m}$ being the stimulated emission probability coupled to the mth mode and $E_{vc,m}$ being the corresponding induced absorption probability, τ_p is the photon lifetime, Γ is the optical confinement factor and $P_{in,m}$ is the rate of injection of signal photons of the mth mode. $F_e(t)$ and $F_{p,m}(t)$ are the fluctuation operators [26] in the form of Langevin shot noise sources to account for fluctuations in n and $N_{p,m}$ caused by random spontaneous emissions [7, 17]. Note that the form of equations (6.28) and (6.29) are similar to the rate equations described in subsection 5.3.1 (see Chapter 5) except for the Langevin noise terms which account for fluctuations caused by spontaneous emissions [27]. The above equations can be solved by first linearising them as [8–28]

$$n(t) = \hat{n} + \Delta n(t) \tag{6.30}$$

$$N_{p,m}(t) = \hat{N}_{p,m} + \Delta N_{p,m}(t) \tag{6.31}$$

In the above equations a circumflex (ˆ) notation denotes the unperturbed steady-state values for the parameters. Expressing R, G_m and $E_{cv,m}$ using these perturbed quantities with Taylor's theorem in equations (6.28) and (6.29) [8]:

$$\frac{d}{dt}\Delta n(t) = \left(-\frac{\partial \hat{R}}{\partial \hat{n}}\sum_m \frac{\partial \hat{G}_m}{\partial \hat{n}} - \Gamma \hat{N}_{p,m}\right) \cdot \Delta n(t) - \sum_m \Gamma \hat{G}_m \Delta N_{p,m}(t) + F_e(t)$$

(6.32)

$$\frac{d}{dt}\Delta N_{p,m}(t) = \Gamma\left(\hat{N}_{p,m}\frac{\partial \hat{G}_m}{\partial \hat{n}} + \frac{\partial \hat{E}_{cv,m}}{\partial \hat{n}}\right)\Delta n(t) + \left(\Gamma \hat{G}_m - \frac{1}{\tau_p}\right)\Delta N_{p,m}(t) + F_{p,m}(t)$$

(6.33)

The above equations can be solved by Fourier transform. To do this, the statistical properties of the Langevin noise sources first have to be known. This can be found by ensemble averaging equations (6.28) and (6.29) in the time domain [1, 3] such that [8]

$$\langle F_e(t)\rangle = \langle F_{p,m}(t)\rangle = 0$$

(6.34)

$$\langle F_e^2(\omega)\rangle = \Im\langle F_e(t) \cdot F_e(s)\rangle$$
$$= P + \hat{R} + \sum_m [\Gamma \hat{E}_{cv,m}\hat{N}_{p,m} + \Gamma(\hat{E}_{cv,m} - \hat{G}_m)\hat{N}_{p,m}]$$

(6.35)

$$\langle F_{p,m}^2(\omega)\rangle = \Im\langle F_{p,m}(t) \cdot F_{p,m}(s)\rangle$$
$$= \frac{\hat{N}_{p,m}}{\tau_p} + \Gamma \hat{E}_{cv,m}(1 + \hat{N}_{p,m}) + \Gamma(\hat{E}_{cv,m} - \hat{G}_m)\hat{N}_{p,m} + P_{in,m}$$

(6.36)

$$\langle F_e(\omega) \cdot F_{p,m}(\omega)\rangle = \Im\langle F_e(t) \cdot F_{p,m}(s)\rangle$$
$$= -\Gamma[\hat{E}_{cv,m}(1 + \hat{N}_{p,m}) + (\hat{E}_{cv,m} - \hat{G}_m)\hat{N}_{p,m}]$$

(6.37)

where $\Im$ represents the Fourier transform and $\langle AB\rangle$ represents the correlation between the two random quantities A and B [3]. Equations (6.35) and (6.36) describe the power spectrum of the Langevin noise sources $F_e(t)$ and $F_{p,m}(t)$, respectively, whereas that in equation (6.37) represents the cross power spectrum between them. It can be seen from these equations that the postulated Langevin noise terms stem from the fluctuations in the carrier density and photon number due to spontaneous emissions (Figure 6.1). Using equations (6.34)–(6.37), equations (6.32) and (6.33) can be solved in the frequency domain for $\langle \Delta n^2(\omega)\rangle$ and $\langle \Delta N_{p,m}^2(\omega)\rangle$ (only the mean-square values can be found because Δn and $\Delta N_{p,m}$ will fluctuate with time but are statistically stationary [1]). The net noise power detected by a photo-detector is related to the total photon number fluctuation spectrum $\langle \Delta N_p^2(\omega)\rangle$, which can be found to consist of two components, one due to self-correlation of the photon number fluctuation in the mth mode $\langle \Delta N_{p,m}(\omega) \cdot \Delta N_{p,m}^*(\omega)\rangle$ and the other due to cross correlation between the kth and lth modes $\langle \Delta N_{p,k}(\omega) \cdot \Delta N_{p,l}^*(\omega)\rangle$ [8]. This can be found by solving equations (6.32) and (6.33) for each mode m and summed over all the possible modes:

$$\langle \Delta N_p^2\rangle = \text{Re}\left[\sum_k\sum_l \langle \Delta N_{p,k}(\omega) \cdot \Delta N_{p,l}^*(\omega)\rangle\right]$$

(6.38)

taking both components into consideration. The resulting noise power at the detector with load resistance R_L and bandwidth B_0 with perfect coupling is given by [16]

$$P_0(\omega) = \frac{\langle \Delta N_p^2(\omega) \rangle}{\tau_{p2}^2} R_L B_0 \eta_D^2 e^2 \qquad (6.39)$$

where η_D is the quantum efficiency of the photo-detector and τ_{p2} is the photon lifetime due to output mirror loss (see subsection 5.2.1 of Chapter 5). Compared with the photon statistics approach, the rate equation analysis is more complicated. In addition, only the total noise power can be derived using equation (6.38) and the corresponding noise components have to be calculated by using more indirect methods [8]. However, it is better than the photon statistics formulation in the sense that we can take into account more factors (e.g. saturation of the amplifier gain due to finite population inversion: see subsections 2.3.3 and 5.2.2 and Section 5.3) by this complex modelling. It is also interesting to note that Mukai and Yamamoto [8] discovered that the rate equation and photon statistics analysis give slightly different theoretical results. Experimental results tend to lie between the theoretical limits imposed by these two methods. The difference arises because of the difference in formulating the signal gain G. In the photon statistics analysis we calculate G by using an active FP formulation, whereas the rate equation analysis requires a gain in the format of equation (5.17).

6.2.3 *Travelling-wave equations formulation*

Both the rate equation and photon statistic master equation approaches ignore the spatial dependence of the signal and noise photons within the amplifier. In order to obtain a more complete picture of the effects of different sources of noise, as well as their effects on the spectral content of the amplifier output, Hinton [22–24] modified the travelling-wave equations that describe SLAs (subsection 5.3.2). He introduced Langevin noise terms in the travelling-wave equations for the amplitude and phase of the output signal, in addition to $F_e(t)$. The results are very complex and apparently neither analytical nor numerical solutions are available so far. Nevertheless, he was able to predict linewidth broadening due to spontaneous emissions, which was followed by the studies carried out by Kikuchi *et al.* [5] using a similar but simpler analysis. Because of the complexity of this approach, we will not pursue this formation in any further detail.

6.3 ANALYSIS OF NOISE IN SLAs USING THE EQUIVALENT CIRCUIT MODEL

The discussion in the preceding section has illustrated the basics of analysing noise in SLAs. It can be seen that they all involve the analysis of spontaneous emissions from SLAs, and the physical processes in semiconductor lasers have to be understood clearly before such analysis can be made with the aid of probability and statistical theory. In Section 5.5 (see Chapter 5) an equivalent circuit model for a SLA was developed. We saw how this model can enable us to analyse the gain and saturation characteristics efficiently. In this section we will see how this equivalent circuit model can help us to understand and analyse spontaneous emissions and the resulting noise characteristics of SLAs. This is because we can use circuit analogies, familiar to most engineers, to formulate the noise processes in the equivalent circuit instead of involving specialised theories like those used in Section 6.2. The representation of spontaneous emissions in a SLA by a circuit model will be examined first.

6.3.1 Representation of spontaneous emissions in a SLA by an equivalent circuit

Recall that in the equivalent circuit model the SLA is represented by a cascade of M sub-networks, each consisting of a negative resistance (to account for the net optical gain) and a small section of lossless transmission line of length $\Delta z = L/M$, with L being the length of the amplifier. For the kth sub-network (corresponding to the kth section) used in transfer matrix analysis (see subsection 5.4.2 in Chapter 5) the negative resistance is denoted by Z_k. For a single-cavity SLA, the characteristics impedance Z_0 of all the lossless transmission lines in each sub-network is the same and can be used to model real optical gain in the kth section of the SLA by the following analogy (Section 5.5, Chapter 5):

$$\frac{Z_0}{Z_0 + Z_k} = \exp\left(\frac{g_k \Delta z}{2}\right) \tag{6.40}$$

if Δz is small enough. Then

$$\frac{Z_0}{Z_k} \approx -\frac{g_k \Delta z}{2} \tag{6.41}$$

Since the major properties of a SLA, such as optical gain, can be modelled by their electrical circuit analogue in this equivalent circuit model, it is natural to also seek an electrical analogue for spontaneous emissions. The similarity between spontaneous emissions in semiconductor lasers and shot noise in electrical circuits has been discussed in Section 6.1. Thus an analogy between them can be established. Consider the case when there is no signal incident onto the SLA. In the equivalent circuit model this means that there is no signal voltage or current flowing. Then the only possible shot noise voltage source in the circuit will be due to thermal fluctuations of the lumped impedances. These thermal shot noises were analysed by Nyquist using the equipartition theorem of statistical mechanics [16]. The single-sided power spectral density (PSD) [3] of this noise voltage in a load resistance R_L is given by [16, 29]:

$$S_{th} = 4kTR_L \tag{6.42}$$

where k is the Boltzmann constant and T is the temperature of the resistor.

Louisell [30] showed that in the region where the quantum energy $\hbar\omega$ is comparable with the thermal energy kT, the *zero-point fluctuations* owing to these quanta are also important. This can be analysed using quantum mechanics, and the corresponding noise has a *single-sided* PSD S_q given by [29]

$$S_q = 2\hbar\omega R_L \qquad (6.43)$$

If we are using the equivalent circuit to model optical frequencies, then the quantity $\hbar\omega \gg kT$ and hence the contribution of thermal fluctuation to the total noise voltage is negligible [29]. Thus, in such circumstances the PDS (single-sided) of the noise voltage in a resistance R_L which is due to quantum mechanical fluctuations is given by

$$S_e \approx S_q = 2\hbar\omega R_L \qquad (6.44)$$

The above formula is applicable for a passive resistance R_L. For a negative resistance, Bjork and Nilsson [44] suggested that equation (6.44) can be applied to that for a negative resistance Z if R_L is replaced by the magnitude of Z, that is

$$S_e = 2\hbar\omega|Z| \qquad (6.45)$$

The above postulation can be justified by the following argument. Although a passive resistance R_L absorbs electrical power through dissipation, whereas a negative resistance Z releases power, the quantum mechanical fluctuations in both types of resistance are shot noise voltage sources (or current sources depending on the way of modelling the noise source [3]) which will act as a power source as well. The amount of fluctuation in both types of resistance depends on the *magnitude* of the power absorbed (in passive resistance) or released (in negative resistance). Hence one can use the *magnitude* of the resistance to calculate the power spectrum of the quantum mechanical fluctuations in them, as in equations (6.44) and (6.45).

We have now completed the mathematical representation of quantum mechanical fluctuations in the equivalent circuit of the SLA. In the optical frequency regime, which is the frequency range for our modelling parameters and of primary interest, the thermal shot noise can be neglected. Instead, the major noise sources are due to quantum mechanical fluctuations. These fluctuations arise because the magnitude of the energy quanta $\hbar\omega$ are significant and the statistical uncertainty will also be significant in the measuring or observation process. In the absence of any signal voltage, these zero-point fluctuations will result in a finite reading on a root-mean-square (r.m.s.) voltmeter connected across the output of the circuit [30]. If these observations are compared with a real SLA, biased to give a finite gain, but without any incident optical signal, a strong similarity between them can be observed. In the case of a SLA, because of the population inversion created by the injection current, spontaneous emission will occur (see equation (2.5) in Chapter 2). The amount of spontaneous emissions will depend on the level of population inversion, similar to the fact that quantum mechanical fluctuation in a resistor is proportional to the magnitude of its resistance. In the absence of any signal, the output of a biased SLA will be delivering a finite optical power (see Chapter 7) similar to the recording of zero-point fluctuations at the output of the equivalent circuit described above.

Thus it appears that in order to model spontaneous emissions in SLAs, we should use the equivalent circuit with an added noise voltage source $e_k(t)$ with a single-sided PSD of S_{ek} for each of the sub-networks. The quantity Z in equation (6.45) will be replaced by Z_k, which is defined by analogy to equations (6.40) and (6.41). Notice that *only* the active negative resistance is noisy in each sub-network. The transmission lines and the terminating loads are noiseless. Although it is suggested that spontaneous emissions can be modelled in this manner, because of the similarity between the shot noise in electrical circuits and spontaneous emissions in SLAs [29], this argument does not justify the mathematical use of the PSD of spontaneous emission in the case of SLAs. This can be tested by using the proposed model to analyse the spontaneous emission power from a SLA with a uniform material gain profile and comparing the results with analytical ones which are well documented in the literature (e.g. see Section 6.5 [17]).

6.3.2 Validity of modeling spontaneous emissions by an equivalent circuit

In this subsection the representation of spontaneous emissions in a SLA by the quantum mechanical fluctuations in an equivalent circuit will be used to calculate the spontaneous emission power in the output of a SLA in the absence of an input signal. This information is extremely useful in the analysis of SLAs and has already been used in the photon statistics master equation in subsection 6.2.1 (i.e. equation (6.14)). For SLAs with complex structures, their noise properties can still be analysed following the procedures discussed in subsection 6.2.1, provided the power spectral density of the spontaneous emissions (and hence the photon number spectral density $N_p(f)$ in equation (6.14)) for these structures can be found. As we shall see in this subsection, this can be done easily with the equivalent circuit model. Conventionally, spontaneous emission power from a SLA can be analysed using laser physics [31, 32]. However, the analysis can only be applied to closed optical cavities and will not be valid for open resonator structures such as distributed feedback (DFB) lasers and optical amplifiers. Henry [17] proposed the use of a Green function approach to solve the spontaneous emission fields via a semi-classical approach. The methodology outlined in his paper can be used to analyse DFB lasers and optical amplifiers. In fact, his method was powerful enough to analyse even more complex structures such as coupled cavities. However, finding the Green functions of complex optical structures is not an easy task [33]. This is illustrated by the recent work of Kahen [34]. The situation, interestingly, is quite similar to that in solving the travelling-wave equations (i.e. equations (5.30) and (5.31) discussed in subsection 5.5.2 in Chapter 5). The use of the equivalent circuit model together with the transfer method (TMM) has been proposed by Chu and Ghafouri-Shiraz in Chapter 5 to overcome the

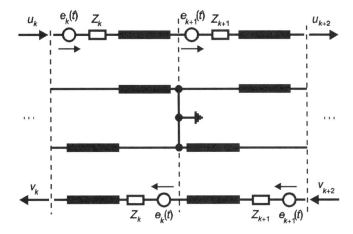

Figure 6.2 Representation of spontaneous emissions in SLAs by the equivalent circuit model

problem encountered with solving travelling-wave equations for optical cavities [56, 57]. It is therefore logical to examine whether it is also possible to analyse spontaneous emissions in SLAs by this technique. In this subsection the feasibility of using this technique will be demonstrated by analysing simple FPAs for which results are well known. Alternative methods of arriving at the same results using Green functions are outlined in Section 6.5.

Consider a FPA with a uniform material gain profile g_m (i.e. $g_{mk} = g_m$ for all k's) and length L. The group effective refractive index of the cavity is N_g, and the power reflection coefficients at the facets are R_1 and R_2, respectively. The equivalent circuit model for this structure is shown in Figure 6.2 together with the proposed noise sources $e_k(t)$ added to each kth sub-network. Consider the kth section in the cavity, which is modelled by the kth sub-network in the equivalent circuit if Figure 6.2. The values of Z_k and Z_0 of the sub-network are given by equations (6.40) and (6.41). In addition, Z_0 is determined by the group velocity of the fields travelling across the circuit, which is related to N_g. The modal gain coefficient g_k at the kth section for a simple buried heterostructure (BH) is given by

$$g_k = g = \Gamma(g_m - \alpha_a) - (1 - \Gamma)\alpha_c \qquad (6.46)$$

where Γ is the optical confinement factor of the active region, and α_a and α_c are the losses in the active region and claddings, respectively. Note that the subscript k in equation (6.46) has been dropped as we have assumed that the material gain profile in the FPA is uniform. Moreover, equation (6.46) has assumed that a unique value of Γ can be defined. However, as discussed in Section 4.3 in Chapter 4, a unique Γ cannot be defined unambiguously in some SLA structures [35]. Then g has to be calculated by the technique introduced in subsection 4.3.1 without using equation (6.46). Following the notation used in subsection 5.4.3 the voltages on both sides of the kth sub-network are related by the following transfer matrix expression (see section 5.5) [56, 57]:

$$\begin{bmatrix} u_k \\ v_k \end{bmatrix} = \begin{bmatrix} \exp(-g_k\Delta z/2)\exp(+j\beta_z\Delta z) & 0 \\ 0 & \exp(g_k\Delta z/2)\exp(-j\beta_z\Delta z) \end{bmatrix} \cdot \begin{bmatrix} u_{k+1} \\ v_{k+1} \end{bmatrix}$$

(6.47)

where the same notation has been used to denote the fields in the SLA and the voltages in the equivalent circuit. Consider the voltages in the equivalent circuit. In the absence of both the signal and spontaneous emissions, $v_k = u_k = 0$ throughout the entire FPA. However, as discussed in subsection 6.3.1, there are finite zero-point fluctuations due to quantum mechanical fluctuations in the resistors Z_k in each sub-network. It remains for us to find the total noise power across the terminating load R_{L2} due to all of these fluctuations in each of the sub-networks in the equivalent circuit. This result will be compared with the output spontaneous emission power at R_2 for a FPA (see Section 6.5). The validity of the equivalent circuit model for analysing noise in SLAs can then be checked. Consider the input voltage u_k to the transmission line of the kth sub-network travelling in the $+z$-direction. This voltage will have two components: one due to fluctuations in Z_k of the sub-network itself, and the other accumulated fluctuations generated and amplified by all the previous $(k-1)$ sub-networks. We denote the former component by Δu_k and the latter component by u_k^{ac} ("ac" represents accumulated). Similar notation will be applied to the voltages in the $-z$-direction (i.e. the v_k's) as well. Let us consider the voltages at the termination load where $z = L = M\Delta z$ (which models the left facet of the FPA). Using equation (5.61) in Chapter 5, we arrive at the following expression:

$$\begin{bmatrix} u_{M+1} \\ v_{M+1} \end{bmatrix} = \frac{1}{t_2}\begin{bmatrix} 1 & r_2 \\ r_2 & 1 \end{bmatrix} \cdot \begin{bmatrix} u_{M+2} \\ v_{M+2} \end{bmatrix}$$

(6.48)

where $r_2 = \sqrt{R_1}$ and is given by

$$r_2 = \frac{R_{L2} - Z_0}{R_{L2} + Z_0}$$

(6.49)

u_{M+1} and v_{M+1} are given by

$$u_{M+1} = u_{M+1}^{ac} + \Delta u_{M+1}$$

(6.50a)

$$v_{M+1} = v_{M+1}^{ac} + \Delta v_{M+1}$$

(6.50b)

Examining the meaning of each term in equations (6.50a) and (6.50b), it can be seen that

$$\begin{bmatrix} u_{M+1}^{ac} \\ v_{M+1}^{ac} \end{bmatrix} = \begin{bmatrix} \exp(g_M\Delta z/2)\exp(-j\beta_z\Delta z) & 0 \\ 0 & \exp(-g_M\Delta z/2)\exp(j\beta_z\Delta z) \end{bmatrix} \cdot \begin{bmatrix} u_M \\ v_M \end{bmatrix}$$

(6.51)

Note that equation (6.51) has been rearranged based on equation (6.47). (Notice the difference in the indices in the column vectors in the left-hand side and the right-hand side of equation (6.51).) It can be seen that in general

$$\begin{bmatrix} u_{k+1}^{ac} \\ v_{k+1}^{ac} \end{bmatrix} = \begin{bmatrix} \exp(g_k\Delta z/2)\exp(-j\beta_z\Delta z) & 0 \\ 0 & \exp(-g_k\Delta z/2)\exp(j\beta_z\Delta z) \end{bmatrix} \cdot \begin{bmatrix} u_k \\ v_k \end{bmatrix}$$

(6.52)

where

$$u_k = u_k^{ac} + \Delta u_k \tag{6.53a}$$

$$v_k = v_k^{ac} + \Delta v_k \tag{6.53b}$$

The Δ terms in equations (6.53a) and (6.53b) are fluctuations generated by the kth sub-network. Equations (6.52) and (6.53) can be used iteratively to evaluate u_{M+1} and v_{M+1} in equations (6.51) and (6.55). Using equation (6.51) in equation (6.50a) we arrive at the following expression:

$$u_{M+1} = u_{M+1}^{ac} = \sqrt{A_M}\,(u_M^{ac} + \Delta u_M)\exp(-j\beta_z\Delta z) \tag{6.54}$$

where the first term on the right-hand side of equation (6.54) is u_M^{ac} given by equation (6.51). The term $\sqrt{A_M}$ is used to represent the term $\exp(g_m\Delta z/2)$ for simplicity. Note that $\Delta u_{M+1} = \Delta v_{M+1} = 0$ as the terminating load R_{L2} is noiseless. However, as shown by equation (6.52):

$$u_M^{ac} = \sqrt{A_{M-1}}\,(u_{M-1}^{ac} + \Delta u_{M-1})\exp(-j\beta_z\Delta z) \tag{6.55}$$

Then equation (6.54) becomes

$$\begin{aligned} u_{M+1} &= \sqrt{A_M}\{\sqrt{A_{M-1}}\,\exp(-i\beta_z\Delta z)[u_{M-1}^{ac} + \Delta u_{M-1}] + \Delta u_M\}\exp(-j\beta_z\Delta z) \\ &= P_M[P_{M-1}u_{M-1}^{ac} + P_{M-1}\Delta u_{M-1} + \Delta u_M] \end{aligned} \tag{6.56}$$

where we have used

$$P_M = \sqrt{A_M}\exp(-j\beta_z\Delta z) \tag{6.57}$$

But from equation (6.52) we have

$$u_{M-1}^{ac} = \sqrt{A_{M-2}}\,(u_{M-2}^{ac} + \Delta u_{M-2})\exp(-j\beta_z\Delta z) \tag{6.58}$$

Substituting this into equation (6.56) gives

$$u_{M+1} = P_M[P_{M-1}P_{M-2}(u_{M-2}^{ac} + \Delta u_{M-2}) + P_{M-1}\Delta u_{M-1} + \Delta u_M] \tag{6.59}$$

If we carry on the iteration with equation (6.52), we finally arrive at

$$u_{M+1} = P_M\left[P_{M-1}P_{M-2}\cdots P_1 u_1^{ac} + \sum_{r=1}^{M-1}\left(\prod_{k=r}^{M-1} P_k\right)\Delta u_r + \Delta u_M\right] \tag{6.60}$$

u_1^{ac} can be found from equation (5.51) in Chapter 5 at $z = 0$:

$$\begin{bmatrix} u_0 \\ v_0 \end{bmatrix} = \frac{1}{t_1}\begin{bmatrix} 1 & -r_1 \\ -r_1 & 1 \end{bmatrix} \cdot \begin{bmatrix} u_1^{ac} \\ v_1^{ac} \end{bmatrix} \tag{6.61}$$

where $r_1 = \sqrt{R_1}$ and is given by

$$r_1 = \frac{R_{L1} - Z_0}{R_{L1} + Z_0} \tag{6.62}$$

Notice that $\Delta u_0 = \Delta v_0 = 0$ as R_{L1} is considered to be noiseless. In addition, as there is no signal coming into the amplifier, $u_0 = 0$. From equation (6.61), it can be shown that

$$u_1^{\text{ac}} = r_1 v_1^{\text{ac}} = r_1 v_1 \tag{6.63}$$

Equation (6.60) can then be re-written as

$$u_{M+1} = P_M \left[r_1 P_{M-1} P_{M-2} \cdots P_1 v_1 + \sum_{r=1}^{M-1} \left(\prod_{k=r}^{M-1} P_k \right) \Delta v_r + \Delta v_M \right] \tag{6.64}$$

By following similar procedures in the reverse direction for v_k, it can be shown that v_1 takes a similar form as to equation (6.60), that is

$$v_1 = v_1^{\text{ac}} = P_1 \left[P_2 P_3 \cdots P_M u_{M+1}^{\text{ac}} + \sum_{r=2}^{M} \left(\prod_{k=r}^{M} P_k \right) \Delta u_{M+3-r} + \Delta u_2 \right] \tag{6.65}$$

If the amplifier under analysis has a uniform material gain profile, then $P_1 = P_2 = \cdots = P_M = P$. Equations (6.64) and (6.65) then become:

$$u_{M+1} = P[P^{M-1} u_1^{\text{ac}} + P^{M-1} \Delta u_1 + P^{M-2} \Delta u_2 + \cdots + P \Delta u_{M-1} + \Delta u_M] \tag{6.66}$$

$$v_1 = P[P^{M-1} v_{M+1}^{\text{ac}} + P^{M-1} \Delta v_{M+1} + P^{M-2} \Delta v_M + \cdots + P \Delta v_3 + \Delta v_2] \tag{6.67}$$

Using equation (6.64) in equations (6.66) and (6.67) we obtain the following expression:

$$u_{M+1} = P\{r_1 P^{M-1} P[P^{M-1} v_{M+1}^{\text{ac}} + P^{M-1} \Delta v_{M+1} + P^{M-2} \Delta v_M + \cdots + P \Delta v_3 + \Delta v_2] \\ + [P^{M-1} \Delta u_1 + P^{M-2} \Delta u_2 + \cdots + P \Delta u_{M-1} + \Delta u_M]\} \tag{6.68}$$

But from equation (6.48) we have

$$v_{M+1}^{\text{ac}} = r_2 u_{M+1}^{\text{ac}} = r_2 u_{M+1} \tag{6.69}$$

as $v_{M+2} = 0$ with no input signal to the FPA. Substituting equation (6.69) into equation (6.68), we obtain:

$$u_{M+1} = P\{r_1 r_2 P^{2M-1} \cdot u_{M+1} + [r_1 P^{2M-1} \Delta v_{M+1} + r_1 P^{2M-2} \Delta v_M + \cdots + r_1 P^{M+1} \Delta v_3 \\ + r_1 P^M \Delta v_2] + [P^{M-1} \Delta u_1 + P^{M-2} \Delta u_2 + \cdots + P \Delta u_{M-1} + \Delta u_M]\} \tag{6.70}$$

Rearranging equation (6.70) by grouping the terms for u_{M+1} by using equation (6.48) again for u_{M+2}, we finally arrive at the following expression:

$$u_{M+1} = t_2 u_{M+1}$$
$$= \left(\frac{t_2 P}{1 - r_1 r_2 P^{2M}} \right) \cdot [r_1 (P^{2M-1} \Delta v_{M+1} + P^{2M-2} \Delta v_M + \cdots + P^M \Delta v_2) \\ + (P^{M-1} \Delta u_1 + P^{M-2} \Delta u_2 + \cdots + \Delta u_M)] \tag{6.71}$$

The mean-square value of the voltage u_{M+1} can be found as

$$
\begin{aligned}
\langle |u_{M+2}|\rangle^2 &= \frac{(1-R_2)|P|^2}{|1-r_1 r_2 P^{2M}|^2} \cdot \{R_1|P|^{2M}[\langle|\Delta v_2|^2\rangle + |P|^2\langle|\Delta v_3|^2\rangle + \cdots \\
&\quad + |P|^{2M-2}\langle|\Delta v_{M+1}|^2\rangle + (|P|^{2M-2}\langle|\Delta u_1|^2\rangle + |P|^{2M-4}\langle|\Delta u_2|^2\rangle + \cdots \\
&\quad + \langle|\Delta u_M|^2\rangle)\}
\end{aligned}
\tag{6.72}
$$

where we have assumed that the fluctuations have no spatial correlation, i.e. $\langle\Delta v_i\Delta v_j^*\rangle = 0$ and $\langle\Delta u_i\Delta u_j^*\rangle = 0$ for $i\neq j$. In addition, there is no such correlation for voltages travelling in opposite directions, i.e. $\langle\Delta u_i\Delta v_j^*\rangle = 0$. But $\langle|u_k|^2\rangle = \langle|v_k|^2\rangle = \langle|e(t)|^2\rangle/(W d Z_0)$; recall that we have normalised u_k and v_k with respect to the optical intensity. If the power of the total fluctuation is measured at $z = L$ in a small optical bandwidth $\Delta\omega = 2\pi f$, then by Fourier analysis it can be shown that [3]

$$
\langle|e^2(t)|\rangle = \frac{S_e\Delta\omega}{2\pi} \tag{6.73}
$$

Then the power P_0 measured at the output of the equivalent circuit will be given by

$$
P_0 = \frac{\Delta\omega}{2\pi Z_0} \cdot S_e \cdot \frac{(1-R_2)|P|^2}{|1-r_1 r_2 P^{2M}|^2} \cdot (R_1|P|^{2M}+1)\sum_{r=0}^{M-1} P^{2r} \tag{6.74}
$$

For a FPA with a uniform material gain profile:

$$
P^{2M} = \exp(g M\Delta z)\exp(-2j\beta_z M\Delta z) = \exp(gL)\exp(-2j\beta_z L) \tag{6.75}
$$

where $L = M\Delta z$. Then equation (6.74) can be re-written as

$$
P_0 = \frac{\Delta\omega}{2\pi Z_0} \cdot S_e \cdot \frac{G_{FP}|P|^2}{(1-R_1)|P|^{2M}} (R_1|P|^{2M}+1)\left(\frac{|P|^{2M}-1}{|P|^2-1}\right) \tag{6.76}
$$

where we have used the sum of a geometric series to evaluate ΣP^{2r}. The term G_{FP} is the gain of a FPA with uniform gain profile and is given by equation (5.4) of subsection 5.2.1 in Chapter 5, that is:

$$
G_{FP} = \frac{(1-R_1)(1-R_2)|P|^{2M}}{|1-\sqrt{R_1 R_2}\,P^{2M}|^2} = \frac{(1-R_1)(1-R_2)\exp(gL)}{|1-\sqrt{R_1 R_2}\,\exp(gL)\exp(-2j\beta_z L)|^2} \tag{6.77}
$$

Since $|P|^2 = \exp(g\Delta z) = A$ and $|P|^{2M} = \exp(g M\Delta z) = \exp(gL)$, then equation (6.76) can be re-written as

$$
P_0 = \frac{\Delta\omega}{2\pi Z_0} \cdot S_e \cdot \left(\frac{A}{A-1}\right) \cdot \frac{G_{FP}}{(1-R_1)} (R_1 e^{gL}+1)(1-e^{-gL}) \tag{6.78}
$$

If Δz is small enough, A can be expanded by Taylor's series. Then it can be shown that

$$
\frac{A-1}{A} = 1-\frac{1}{A} = 1-\exp(-g\Delta z) \approx g\Delta z \tag{6.79}
$$

To proceed further we have to examine the fluctuation in Z_k for each sub-network. It has been shown in section 5.5 that Z_k represents the *net* optical gain. This can be split into three components [14]:

$$Z_k = Z_{1k} + Z_{2k} + Z_{ak} \qquad (6.80)$$

where Z_{1k} is a negative resistance modelling the stimulated emissions, Z_{2k} is a passive resistance modelling absorption and Z_{ak} is a passive resistance modelling other optical losses like scattering. Any spontaneous emissions will add photons randomly to the stimulated emissions, and hence the major fluctuations in Z_k are due to Z_{1k} only (note that in modelling laser oscillators, quantum fluctuations in the other processes can be quite significant [14]). Then we can write

$$S_{ek} \approx 2\hbar\omega|Z_{1k}| \qquad (6.81)$$

It has also been shown that g can be approximated by equation (6.41) for small Δz. Combining with equation (6.46) in an index guided structure, it can be seen that

$$\frac{Z_{1k} + Z_{2k} + Z_{ak}}{Z_0} = -\{\Gamma g_{mk} - [\Gamma\alpha_{ak} + (1 - \Gamma)\alpha_{ck}]\}\frac{\Delta z}{2} \qquad (6.82)$$

Then by comparing the analogy between each component of Z_k on the left-hand side of the above equation with that on the right-hand side, we arrive at the following expression:

$$Z_{1k} + Z_{2k} = -\frac{\Gamma g_{mk} Z_0 \Delta z}{2} \qquad (6.83)$$

This can be simplified further by noting that the population inversion parameter n_{sp} can be modelled by [14]

$$n_{sp} = \frac{|Z_{1k}|}{|Z_{1k}| - Z_{2k}} = \frac{Z_{1k}}{Z_{1k} + Z_{2k}} \qquad (6.84)$$

Hence

$$|Z_{1k}| = -\frac{\Gamma g_{mk} n_{sp} Z_0 \Delta z}{2} \qquad (6.85)$$

Substituting into equation (6.81) we arrive at:

$$S_{ek} = \hbar\omega\Gamma g_{mk} n_{sp} Z_0 \Delta z \qquad (6.86)$$

The above equation is actually very interesting. It can be seen from Section 5.7 of Chapter 5 that by using the expressions for β and R_{sp} derived there, the term S_{ek} in the above equation actually will be equal to the number of spontaneous emission photons coupled to the signal mode in the optical cavity (i.e. $\beta R_{sp} N_g \Delta z/c$) *provided that* $\Delta\omega = c/(N_g\Delta z)$. Such a correction factor for the spontaneous emission factor β has also been found by Sommers [36], which has been attributed to the difference in using the frequency independent photon statistical interpretation of β whilst analysing the photons in the cavity which are pro-rated over a frequency range. When the equivalent circuit model is used, the amount of spontaneous emission generated at each section is controlled by Δz and g, which is what the above equation implies. Thus one can actually "fine-tune" the values of Δz such that the correction factors in the rate equations applied over larger optical cavities will not be necessary, and a coherent interpretation of the physical processes described by the equivalent circuit model can be obtained [36]. The form of equation (6.86) suggests that the power spectrum of the fluctuations induced by spontaneous emissions in a SLA depends on g_{mk}, the material gain coefficient. In practice, because the material gain coefficient depends on the emission wavelength (see subsection 3.2.2 in Chapter 3), it follows that the power spectrum of the spontaneous emissions will also depend on the shape of the material gain spectrum [37–39]. Substituting this into equation (6.78) for a FPA with a uniform gain profile, we finally arrive at the following expression:

$$P_0 = \frac{\Delta\omega}{2\pi}\,\hbar\omega\left(\frac{n_{sp}}{\eta}\right)\left(\frac{R_1 e^{gL}+1}{1-R_1}\right)(1-e^{-gL})G_{FP} \tag{6.87}$$

where η is the quantum efficiency of the laser amplifier given by [14, 17]:

$$\eta = \frac{g}{\Gamma g_m} \tag{6.88}$$

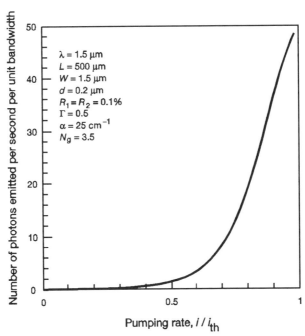

Figure 6.3 Variation of the number of photons emitted per second per unit bandwidth due to spontaneous emissions versus different pumping rates for a FPA with the parameters shown in the inset

for an amplifier with a uniform gain profile. If we compare equation (6.87) with the expression for P_0 derived using Green functions in Section 6.5, it can be seen that they are identical. Hence, by adding the noise voltages $e_k(t)$ at each kth sub-network in the equivalent circuit, with a magnitude given by equations (6.73) and (6.81), the equivalent circuit can also be used to model spontaneous emissions accurately. In Figure 6.3 the variation of $P_0/(\hbar\omega\Delta\omega)$ versus the pumping rate i/i_{th} for a SLA has been plotted. The structural parameters for this SLA are shown in the inset of the figure. It can be seen that the amount of spontaneous emission is not very significant for moderate values of i/i_{th}, and then increases rapidly after i/i_{th} rises beyond 0.7. This supports our proposal to ignore the coupling of spontaneous emissions to the signal mode in the TMM analysis discussed in Chapter 5.

6.3.3 Effects of stray reflections on the spontaneous emission power from a SLA

In this subsection the equivalent circuit model will be used to analyse spontaneous emissions of SLAs subject to reflections from fibre ends [43]. The subsequent effect on the gain characteristics will also be examined. When a SLA has to be used in a system, it has to be coupled to two pieces of optical fibre, one acting as the input and the other as the output. There are discontinuities in the refractive indices between the SLA facet–air and air–fibre interfaces, which will cause reflection [33]. The effects of the reflections on the performance of the SLAs in the system are not well understood, although some work has been done with fibre amplifiers [40, 41]. A formal study of the problem can be performed by the proposed method of transfer matrix analysis and equivalent circuit modelling. Here, the effect of the reflections on the spontaneous emissions from a SLA will be studied. This property is important for multi-amplifier systems because if the spontaneous emissions are enhanced by these stray reflections (as in FPAs where the excess noise coefficient χ comes in, see subsection 6.2.1), then the system may oscillate, or in some SLAs unexpected saturation will be observed [40, 42]. We simplify the analysis by assuming that the reflections from the ends of the optical fibres can be modelled by a plane mirror extending to infinity in both the x and y directions (Figure 6.4(a)) [43]. The power reflection coefficients of these mirrors are R_3

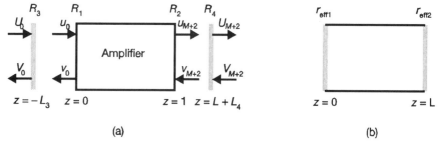

(a)   (b)

Figure 6.4 (a) Configuration used in the analysis of stray reflections from the fibre ends. (b) The equivalent FPA of the configuration in (a)

and R_4, respectively, located at $z = -L_3$ and $z = L + L_4$. In the analysis we assume the amplifier has a uniform gain profile and is divided into M sections. The relationship between U_0, V_0 and u_0, v_0 can be found by the following transfer matrix [44]:

$$\begin{bmatrix} U_0 \\ V_0 \end{bmatrix} = \frac{1}{t_3} \begin{bmatrix} 1 & -r_3 \\ -r_3 & 1 \end{bmatrix} \begin{bmatrix} \exp(jk_0 L_3) & 0 \\ 0 & \exp(-jk_0 L_3) \end{bmatrix} \begin{bmatrix} u_0 \\ v_0 \end{bmatrix} \tag{6.89}$$

Similarly

$$\begin{bmatrix} u_{M+2} \\ v_{M+2} \end{bmatrix} = \frac{1}{t_4} \begin{bmatrix} \exp(jk_0 L_4) & 0 \\ 0 & \exp(-jk_0 L_4) \end{bmatrix} \begin{bmatrix} 1 & r_4 \\ r_4 & 1 \end{bmatrix} \begin{bmatrix} U_{M+2} \\ V_{M+2} \end{bmatrix} \tag{6.90}$$

In the above equations, $r_3 = \sqrt{R_3}$ and $r_4 = \sqrt{R_4}$ are the amplitude reflectivities of the mirrors, and t_3 and t_4 are the corresponding transmittivities given by $t_4^2 = 1 - r_4^2$. The parameter k_0 is the wave number in free space which is assumed to be the medium between the mirrors and the facets of the amplifiers. From equations (6.89) and (6.90) it can be shown that

$$u_0 = r_3 v_0 \exp(-\gamma_1 L_3) \tag{6.91a}$$

$$v_{M+2} = r_4 v_{M+2} \exp(-\gamma_2 L_4) \tag{6.91b}$$

Imposing these conditions onto equations (6.48) and (6.61) we arrive at the following relations:

$$\left(\frac{u_1}{v_1} \right) = r_{\text{eff1}} = \frac{r_1 + r_3 \exp(-\gamma_1 L_3)}{1 + r_1 r_2 \exp(-\gamma_1 L_3)} \tag{6.92a}$$

$$\left(\frac{u_{M+1}}{v_{M+1}} \right) = r_{\text{eff2}} = \frac{r_2 + r_4 \exp(-\gamma_2 L_4)}{1 + r_2 r_4 \exp(-\gamma_2 L_4)} \tag{6.92b}$$

where

$$\gamma_1 = 2jk_0 + \alpha_{13} \tag{6.92c}$$

$$\gamma_2 = 2jk_0 + \alpha_{24} \tag{6.92d}$$

In the above equations, $k_0 = 2\pi/\lambda$ is the free space wave number with λ being the wavelength of the light source. α_{13} is the total loss between the mirror at $z = -L_3$ and the amplifier facets at $z = 0$, and α_{24} is the total loss between the mirror at $z = L + L_4$ and the amplifier facet at $z = L$ [43]. Equation (6.92a) will be used in place of equation (6.63), whereas equation (6.92b) will be used in place of equation (6.69). With these replacements and following the procedures outlined in the previous subsection, we arrive at the following expression (see equation (6.71)):

$$u_{M+2} = \left(\frac{t_{\text{eff2}} P}{1 - r_{\text{eff1}} r_{\text{eff2}} P^{2M}} \right) \cdot [r_{\text{eff1}} (P^{2M-1} \Delta v_{M+1} + P^{2M-2} \Delta v_M + \cdots + P^M \Delta v_2)$$
$$+ (P^{M-1} \Delta u_1 + P^{M-2} \Delta u_2 + \cdots + \Delta u_M)] \tag{6.93}$$

where t_{eff2} is the effective transmittance of the configuration given by

$$|t_{\text{eff2}}|^2 = 1 - |r_{\text{eff2}}|^2 \tag{6.94}$$

Other terms have been defined in the previous subsection. An inspection of equation (6.93) reveals that we can replace the composite structures of Figure 6.4(a) by an equivalent FPA in Figure 6.4(b), where the facet reflectivities have been replaced by r_{eff1} and r_{eff2}, respectively. Consequently, the spontaneous emission power P_0 of the structure can be found simply by an equation similar to equation (6.74), namely

$$P_0 = \frac{\Delta\omega}{2\pi Z_0} \cdot S_e \cdot \frac{(1 - |r_{\text{eff2}}|^2)|P|^2}{|1 - r_{\text{eff1}} r_{\text{eff2}} P^{2M}|^2} \cdot (|r_{\text{eff1}}|^2 |P|^{2M} + 1) \sum_{r=0}^{M-1} P^{2r} \tag{6.95}$$

which can be simplified to give

$$P_0 = \frac{\Delta\omega}{4\pi^2} \hbar\omega \left(\frac{n_{\text{sp}}}{\eta}\right) \left(\frac{|r_{\text{eff1}}|^2 \exp(gL) + 1}{1 - |r_{\text{eff1}}|^2}\right)[1 - \exp(-gL)]G'_{\text{FP}} \tag{6.96}$$

where G'_{FP} is given by:

$$G'_{\text{FP}} = \frac{(1 - |r_{\text{eff1}}|^2)(1 - |r_{\text{eff2}}|^2)\exp(gL)}{|1 - r_{\text{eff1}} r_{\text{eff2}} \exp(gL)\exp(-2j\beta_z L)|^2} \tag{6.97}$$

We calculate $P_0/(\hbar\omega\Delta\omega)$ for a FPA using equations (6.96) and (6.97) with $R = R_2 = 0.1\%$, $L = 500\,\mu m$, $\alpha = 25\,cm^{-1}$, $\lambda = 1.5\,\mu m$, $N_g = 3.5$, $W = 1.5\,\mu m$,

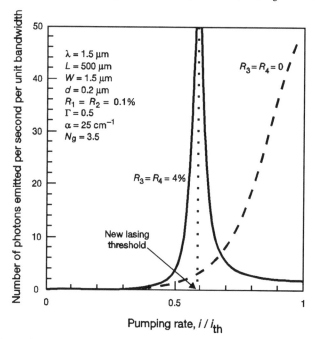

Figure 6.5 Variation of the number of photons emitted per second per unit bandwidth versus the pumping rate for a FPA taking into account the effect of stray reflections from the ends of the optical fibres

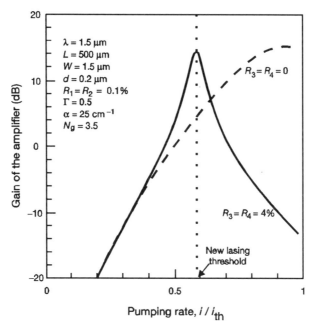

Figure 6.6 Variation of gain versus pumping rate for a FPA taking into account the effect of stray reflections from the ends of the optical fibres

$d = 0.2\,\mu m$ and $\Gamma = 0.5$ versus the pumping rate $\gamma = i/i_{th}$ by considering both the case with fibre end reflections of magnitude 4% and the case where the reflections from the fibre can be ignored. The separation between the fibres and the end facets is assumed to be the same at both the input and the output, and is fixed to $10\,\mu m$. The results are plotted in Figure 6.5. It can be seen that the reflections actually enhance the growth of spontaneous emissions with pumping and the entire structure reaches a lasing threshold earlier than that for a solitary SLA. The actual value of the new lasing threshold current and the amount of enhancement owing to fibre reflections is determined by both the magnitude of the stray reflections as well as the distance between the facets and the fibre ends. Similarly, the gain of the amplifier without an input signal is also enhanced by the stray reflections, as shown in Figure 6.6. These curves can be used to calculate how much the amplifier gain and spontaneous emissions are enhanced by the stray reflections from the fibre ends. Because in most cases an optical fibre transmission system was designed by ignoring such reflections on the noise and gain of the SLAs used as in-line repeaters, these curves will be helpful in showing how the engineer should reduce the pumping to the amplifiers in order to avoid gain saturation or overloading by spontaneous emissions.

To complete the analysis, the effect of stray reflections on the amplifier gain with a finite input signal is examined by using TMM analysis. The parameters used in the analysis are the same as those used in Figures 6.5 and 6.6. The

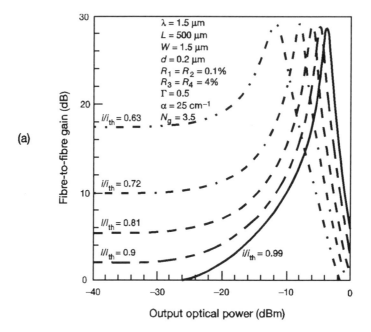

(a)

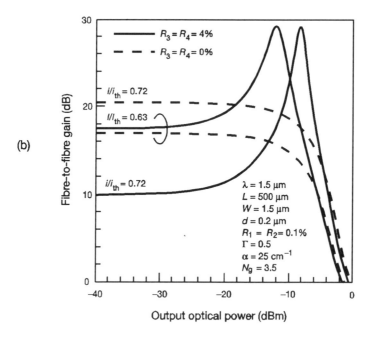

(b)

Figure 6.7 (a) Variation of amplifier gain versus output optical power for a FPA taking into account the effect of reflections from the ends of the optical fibres. (b) Comparison between the variation of amplifier gain versus the output optical power of a FPA with and without taking into account the effect of stray reflections from the ends of the optical fibres

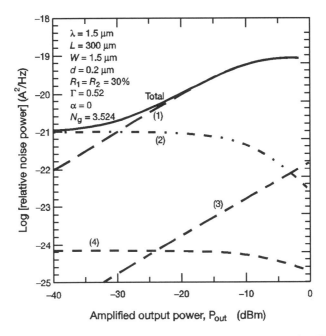

Figure 6.8 Variation of relative noise power generated by a SLA. The different noise components are: (1) signal-spontaneous emission beat noise; (2) beat noise between spontaneous emission components; (3) signal shot noise; (4) spontaneous emission beat noise

results are shown in Figure 6.7(a). Even though the spontaneous emissions coupling to the signal mode are ignored, the calculated results clearly illustrate enhancement of the gain due to the stray reflections which tend to drive the amplifier closer to oscillation. This can be seen more clearly when they are compared with the curves calculated for no fibre end reflections as in Figure 6.7(b), where it can be seen that for a higher output optical power, the effect of fibre reflections becomes more important than at a lower optical power. Finally, it should be noted that in the above analysis the representation of the reflection from the ends of the optical fibres with a plane mirror may not be very accurate in reality. This is especially true for taper-ended fibres or fibres with micro-lenses fabricated on the tip [45, 46]. In these circumstances the effective reflectivities are more complicated than those depicted in equations (6.92a) and (6.92b) (which are actually identical to the functional forms of the reflection coefficient of Fabry–Perot etalons [47]), but an analysis is clearly beyond the scope of the present study.

6.4   APPLICATIONS

We will use the formulation described in the previous sections in two particular ways to examine the effects of the structural parameters of a SLA on its noise

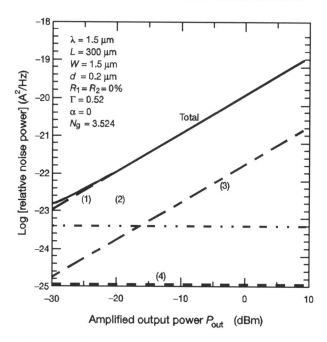

Figure 6.9   Variation of relative noise power generated by a TWA. The different noise components are: (1) signal-spontaneous emission beat noise; (2) beat noise between spontaneous emission components; (3) signal shot noise; (4) spontaneous emission beat noise

characteristics, and to consider the effect of noise in systems. The former is important in designing low-noise SLAs, whereas the latter aspect will affect the signal-to-noise (S/N) ratio at the input of the receiver in a simple intensity modulated (IM) optical fibre communication system.

6.4.1   *Device design criteria*

In subsection 5.5.1 in Chapter 5 we considered how to optimise the structure of a SLA to affect its gain characteristics. The polarisation characteristics have also been examined in subsection 4.3.2 in Chapter 4. In this section, the effects of the structural design of a SLA on its noise characteristics will be analysed. The results should be compared with those in subsections 5.6.1 and 4.3.2, where a more complete picture of the optimum SLA structure can be seen. A FPA with the parameters listed in Table 6.1 has been analysed, using the equations above. In Figure 6.8 the relative noise power per unit bandwidth has been calculated using equation (6.22), assuming a unit detection bandwidth and a unit load resistance for the photodiode, for various output powers, with an unsaturated signal gain G fixed to 20 dB, assuming that the signal wavelength is matched to that of the cavity resonance. The four components of total noise power have been separated as (i)–(iv) in Figure 6.8, assuming no

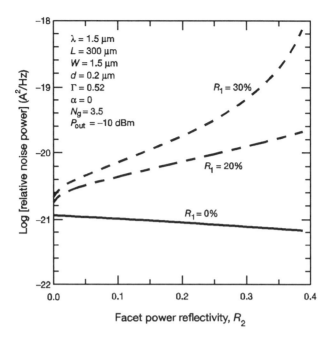

Figure 6.10 Variation of relative noise power generated by a FPA versus R_2, with R_1 as a parameter

Table 6.1 Amplifier parameters used in noise analysis

Unsaturated signal gain	G	20 dB
Facet reflectivities	$R_1 = R_2$	30%
Group index of active region	N_g	3.524
Differential gain coefficient	A	7.2543×10^{-16} cm^{-2}
Transparency carrier density	N_0	1.5034×10^{18} cm^{-3}
Optical confinement factor	Γ	0.52
Amplifier length	L	300 μm
Loss in active region	α	0 cm^{-1}

phase noise in the incident signal. It can be seen that the contribution of the shot noise power is relatively unimportant compared with that of the beat noise, although at higher power levels they can be fairly significant [9]. The beat noise between spontaneous emissions dominates at low power levels. As the signal power increases, the beat noises between the signal and the spontaneous emission become more dominant. At high output power, where gain saturation occurs, the total noise power reduces because the beat noises between the spontaneous emission components are suppressed. This is in accord with the postulate made by Schicketanz and Zeidler [48] in their experiment. Notice that the spontaneous emissions are suppressed when the amplifier is driven towards

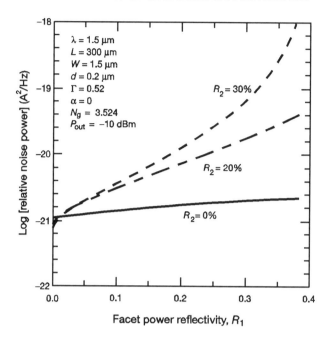

Figure 6.11   Variation of relative noise power generated by a FPA versus R_1, with R_2 as a parameter

saturation. Therefore in Figure 6.8 both shot noise and beat noise between spontaneous emissions reduce with increasing output power.

In Figure 6.9 we repeat the analysis of Figure 6.8 for a TWA with zero values of R_1 and R_2. It can be seen that, on average, the TWA has a better performance compared with the FPA. In Figure 6.9 we do not observe the noise suppression due to gain saturation. This is because the saturation intensity for TWAs is extremely high in the absence of cavity resonance to enhance the saturation mechanism. Hence, the noise level in TWAs can be increased quite substantially by increasing output power, before it is suppressed by saturation mechanisms. In Figures 6.10 and 6.11 we investigate the effects of the reflectivities R_1 and R_2 on the total noise emitted by the amplifier. In Figure 6.10 we have calculated the variation in relative noise power of the amplifier with varying output mirror facet reflectivity R_2 from 0% to 40% for three particular values of the input mirror reflectivity R_1. The output power of the amplifier is fixed to 10 dBm. Our findings are similar to those of Mukai *et al.* [18]. For a transparent input mirror, the noise level reduces with increasing output mirror reflectivity if the output power is maintained. As R_1 increases from 0% to 30%, the noise level increases for a particular value of R_2.

We have performed a similar analysis in Figure 6.11 with R_1 varying from 0% to 40% for three different values of R_2. Unlike Figure 6.10, we observe a cross-over between the curves for low values of R_1. Careful examination of the cross-over reveals that the amplifier can have a lower noise level with $R_1 < 0.05$, but R_2 is finite and high (greater than 20%). Of course the performance is even better where R_2 is zero, in which case the amplifier approaches a nearly travelling-wave amplifier (NTWA) as R_1 is close to zero. Instead of a drop in the noise level for $R_2 = 0$, as with the case of $R_1 = 0$ in Figure 6.10, there is no drop in noise level in all cases in Figure 6.11 as R_1 is increased. In Figure 6.12 we have analysed how the noise figure F can be affected by structural parameters. In Figure 6.12(a) we plot the variation in F for a SLA with equal output and input mirror reflectivities and length L for different combinations of reflectivities and active layer thickness d. The output power is maintained at $-10\,$dBm as L is varied. For the TWA case, where R_1 and R_2 are zero, the active layer thickness has no significant effect on the amplifier noise figure. For $R_1 = R_2 = 0.3$, however, an increase in the active layer thickness reduces the noise figure. In all cases, the noise figure increases with amplifier length, confirming the findings of Saitoh and Mukai [49] that SLAs with short and thick layers tend to have better noise performances. Notice that there are regions where SLAs with finite reflectivities have lower noise levels than TWAs, especially for small L's. Figure 6.12(b) shows the results of an analysis of SLAs with asymmetrical mirrors, i.e. R_1 and R_2 are not equal. In this case, no cross-over of curves is observed. Again, SLAs with thick and short active layers appear to be better in terms of the noise figure. For SLAs with high input mirror reflectivity but low output mirror reflectivity, the noise performance appears to be better than that for SLAs with low input mirror reflectivity but high output mirror reflectivity.

The analysis so far has been based on constant output power. We have analysed the structural effects on the noise figure for a constant density of injected carriers of $1.7 \times 10^{18}\,$cm^{-3} as well as a fixed output power of $-10\,$dBm, i.e. R_{sp} is fixed in calculating the spontaneous emission factor β. This corresponds to the situation where we have tried to maintain the output power by varying the input signal power only. Figure 6.12(c) is the resulting curve with the same structural parameters as in Figure 6.12(a), but under this new assumption. It can be seen that for a constant injection level, F is not affected by the active layer length or thickness for TWAs. On the other hand, there is a dramatic variation in F with length, for SLAs with finite facet reflectivities. For the case where $d = 0.3\,\mu$m, a discontinuity in the curve is observed. This discontinuity represents the structural combinations of L, d and mirror reflectivities with the injection level at which oscillation occurs. A similar observation can be made for the case where $d = 0.15\,\mu$m with R_1 and R_2 being 30%, where the structure approaches oscillation as L approaches $400\,\mu$m. Again, for shorter amplifier lengths the noise performance is better. However, contrary to the condition where output power is fixed, the present calculations show that a thick active layer can actually have poorer noise performance when the injection level is held constant.

In Figure 6.12(d) we perform the same analysis as that for Figure 6.12(b), but with the additional assumption of constant carrier injection. The curves show that in this situation the noise figure exhibits a minimum as L is increased. This minimum corresponds to the best possible achievable noise

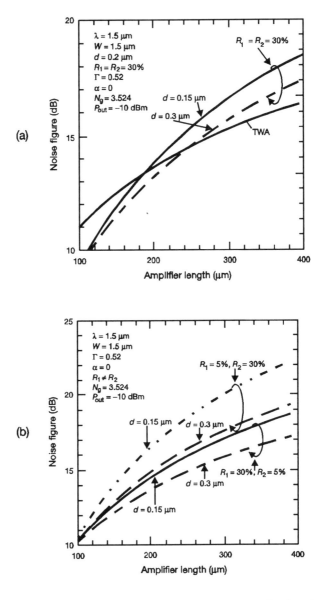

Figure 6.12   (a) Variation of noise figure F versus the amplifier length with the active layer thickness as a parameter. (b) Variation of noise figure versus the amplifier length with thickness as a parameter. (c) Variation of noise figure F versus the amplifier length L with the active layer thickness as a parameter. (d) Variation of noise figure F versus the amplifier length L with the active layer thickness as a parameter

figure for these configurations. Compared with SLAs with identical input and output mirrors, SLAs with asymmetrical mirror reflectivities can be more useful as there is a definite achievable minimum value of the noise figure which is of great help in designing SLA structures. In this case, a thick active layer

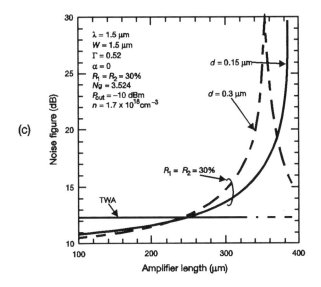

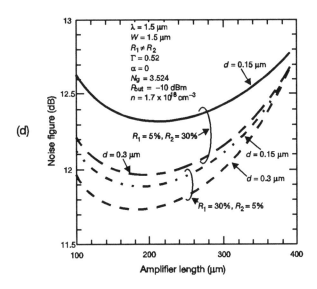

appears to be better in terms of noise performance than the symmetrical case. As with Figure 6.12(b), we found that high input mirror reflectivity but low output mirror reflectivity can help to reduce the noise figure of the SLA. It can be seen from the preceding discussion that the structural effects on the noise figure are extremely complicated. Noise power is affected by operational conditions and hence in design it is necessary to know the operational conditions under which the SLA will be working. In addition, assessment of the best structural combinations is difficult, as they depend strongly on operational conditions.

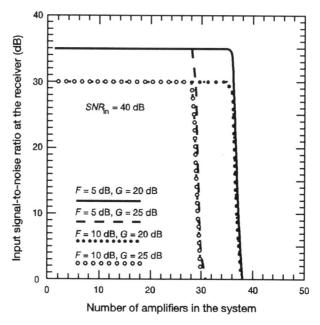

Figure 6.13   Variation of the input signal-to-noise ratio at the optical receiver versus the number of amplifiers in an optical fibre communication link for different values of the amplifier noise figure and signal gain

6.4.2   System considerations

A simple intensity modulation (IM) optical fibre communication system was analysed in subsection 5.6.2 in Chapter 5. There M identical SLAs were used as in-line repeaters, each specified by an amplifier gain G, an output saturation power P_{sat} and a noise figure F. The input signal power from the transmitter to the first optical fibre is P_{in}. Usually, F should depend on G. If $G \gg 1$, then the approximation of equation (6.27) in subsection 6.2.1 applies in the foregoing analysis. For the present study we assume $G = 20\,\text{dB}$ as in subsection 5.6.2. The transmitter will also be emitting noise with a signal-to-noise ratio of $(S/N)_t$. The bit error rate of the receiver with a SLA as its pre-amplifier depends ultimately on the input signal-to-noise ratio of the receiver [3]. Let the minimum input signal-to-noise ratio of the receiver be $(S/N)_{min}$, which means that the receiver can achieve a minimum sensitivity of S [50]. The definition of the noise figure F is given by [3]

$$F = \frac{(S/N)_{in}}{(S/N)_{out}} \tag{6.98}$$

where $(S/N)_{in}$ and $(S/N)_{out}$, respectively, are the input and output signal-to-noise ratios. The signal propagating across a fibre with length L_s kms and with loss α_f/km (see Figure 5.17) will be attenuated by $L_s\alpha_f$. A similar attenuation will be observed for the noise generated by the SLAs. Hence the signal-to-noise ratio for a particular stage within the transmission system is kept constant. Notice that this assumption ignores any phase noise introduced to the signal by Kerr effects along the fibre [51]. Under the above assumption, we can derive the input signal-to-noise ratio to the optical receiver as [7, 11]

$$SNR_d = \frac{(S/N)_t}{F + \dfrac{F}{G} + \dfrac{F}{G^2} + \cdots + \dfrac{F}{G^{M-1}}}$$

(6.99)

It can be seen from the above equation that the noise figure of the first amplifier is the most significant contribution to SNR_d. Hence for cases when different SLAs can be used, the noise figure of the first amplifier in the chain should be a minimum [11]. In Figure 6.13 we have calculated the variation of SNR_d with M for $(S/N)_t = 40$ dB, and G is allowed to be either 20 dB or 25 dB. Similarly, F can be either 5 dB or 10 dB. Interestingly, for the parameters used, the value of the signal-to-noise input to the receiver does not vary much with M until a critical value, where it deteriorates rapidly. This can be explained by the functional form of equation (6.99). Furthermore, it can be seen that increasing G for a particular value of F actually pushes this critical value forward slightly, thus reducing the transmission span. Again, as discussed in Chapter 5, optimum operation of SLAs in a system does not necessarily mean a large value of G. Figure 6.13 should be used together with Figure 5.17 in Chapter 5, because an increase in SNR_0 usually requires an increase in P_{in} (in most cases the noise emitted from the transmitter is difficult to control [52]). This will increase the risk that an amplifier in the chain will become saturated.

6.5 ANALYSIS OF SLA SPONTANEOUS EMISSION POWER USING THE GREEN FUNCTION APPROACH

In Section 6.3 the equivalent circuit model was used to derive the output spontaneous emission power generated by a SLA with a uniform material gain profile. Here we illustrate an alternative derivation for both a TWA and a FPA using Green functions, as first suggested by Henry [17]. Henry considered that the spontaneous emissions in an optical amplifier can be modelled as discrete Langevin noise sources distributed within the cavity. The resulting wave equation becomes an inhomogeneous one with a driving source term. This requires the use of a Green function in finding the solution of the wave equation [33]. In general, the one-dimensional Green function of a one-dimensional wave equation in the frequency domain is given by [53]

$$g_n(z, z') = \frac{Z_{n+}(z_>)Z_{n-}(z_<)}{W_n}$$

(6.100)

where $z_>$ and $z_<$ are the values of z that are greater than and smaller than z', respectively, z' is the coordinate of the source within the structure, and $Z_{n+}(z_>)$ and $Z_{n-}(z_<)$ are the solutions of the nth mode for the homogeneous one-dimensional wave equation travelling in $+z$ and $-z$ directions, respectively. The Wronskian W_n is defined by

$$W_n = Z'_{n+}Z_{n-} - Z_{n+}Z'_{n-} \tag{6.101}$$

where the primes denote derivatives with respect to z. W_n is an independent function of z. The general three-dimensional Green function in the frequency domain is therefore given by

$$G_\omega(r, r') = \sum_n g_n(z, z') \frac{\Phi_n(x, y)\Phi_n(x', y')}{\langle\Phi_n(x, y)\Phi_n(x', y')\rangle} \tag{6.102}$$

where $r = (x, y, z)$ is the position vector in the structure under analysis, and r' is the position vector of the source within the structure. Φ_n is the nth transverse modal field solution, which can be found by solving the homogeneous scalar wave equation for the structure as discussed in Chapter 4. In the following analysis we will use the above expressions to derive the spontaneous emission power generated by TWAs and FPAs with uniform material gain profiles.

6.5.1 Travelling-wave amplifier (TWA)

For an infinite waveguide or a TWA with a perfect anti-reflection coating, the function $Z_{n\pm}$ and W_n can be found as [7]

$$Z_{n\pm} = \exp(\pm j\beta_n z) \tag{6.103}$$

$$W_n = 2j\beta_n \tag{6.104}$$

where β_n is the longitudinal propagation constant of the nth mode, and therefore Green's function for a TWA is given by

$$G_\omega(r, r') = \sum_n \frac{\Phi_n(x, y)\Phi_n(x', y') \exp[j\beta_n|z - z'|]}{2j\beta_n\langle\Phi_n(x, y)\Phi_n(x', y')\rangle} \tag{6.105}$$

The electric field in the structure can be determined from Green's function in the frequency domain by

$$E_\omega(f) = \int dr' G_\omega(r, r')F_\omega(r') \tag{6.106}$$

The subscript ω indicates that the quantities are frequency dependent and the integration is taken over the entire structure. The corresponding magnetic field can be found via Maxwell's equations [33]. The term $F_\omega(r')$ is the magnitude of the Langevin noise source at r'. It satisfies the following statistical properties [7]:

$$\langle F_\omega(r)F_{\omega'}(r')\rangle = \langle F_\omega^*(r)F_{\omega'}^*(r')\rangle = 0 \tag{6.107}$$

$$\langle F_\omega(r)F_{\omega'}^*(r')\rangle = 2D(r)\delta(r - r')\delta(\omega - \omega') \tag{6.108}$$

where D is the diffusion coefficient. It can be shown that by satisfying the quantum statistical properties of radiation, the diffusion coefficient is given by [17, 30]

$$2D(r) = -\frac{\omega^4 h\epsilon''(r)n_{sp}}{\pi c^4} \tag{6.109}$$

where ϵ is the imaginary part of the dielectric constant of the amplifying medium. The noise power within the bandwidth $\Delta\omega$ can be found by first evaluating the Poynting vector and then integrating over the entire transverse section and the frequency spectrum of ω', that is:

$$P_N = \frac{c\Delta\omega}{2\pi} \iint dx \cdot dy \int_0^\infty \langle E_\omega H_{\omega'}^* \exp[-j(\omega - \omega')t] + c \cdot c \rangle d\omega' \tag{6.110}$$

By substituting the diffusion coefficient for the Langevin force term, we obtain the following relation using equations (6.105)–(6.110) above:

$$P_N = \frac{\hbar\omega\Delta\omega g_m n_{sp}\{\exp[(g_m - \alpha_0)L] - 1\}}{2\pi(g_m - \alpha_0)} \tag{6.111}$$

where g_m is the material gain coefficient and α_0 is the loss coefficient. The amplifier is assumed to have length L. Notice that the power amplification factor A is identical to the exponential term in equation (6.111) and since the quantum efficiency of the laser amplifier is given by [14, 54]

$$\eta = \frac{g_m - \alpha_0}{g_m} \tag{6.112}$$

the spontaneous emission noise power is given by

$$P_N = \frac{\Delta\omega\hbar\omega n_{sp}(A - 1)}{2\pi\eta} \tag{6.113}$$

This is a well-known equation, except for the introduction of the quantum efficiency in the denominator. Lowery [55] verified that such an inclusion is necessary to obtain the correct ASE power by TLLM simulation.

6.5.2 Fabry–Perot amplifiers

Consider a FP cavity of length L, extending from $z = 0$ to L. The field reflectivities at the two end facets are given by r_1 and r_2 at $z = 0$ and $z = L$, respectively. Then by considering a single transverse mode operation, the quantities $Z_{n\pm}$ and W_n for $n = 0$ are given as

$$Z_{0-} = r_1 \exp(j\beta_0 z) + r_2 \exp(-j\beta_0 z) \tag{6.114}$$

$$Z_{0+} = r_1[\exp(j\beta_0 z) + r_2 \exp(2j\beta_0 L - j\beta_0 z)] \tag{6.115}$$

which satisfies the boundary condition. The additional factor r_1 in equation (6.115) is there so that at the threshold, $Z_{0+} = Z_{0-}$. The Wronskian is given by

$$W_0 = 2j\beta_0 r_1[1 - r_1 r_2 \exp(2j\beta_0 L)] \tag{6.116}$$

Substituting these expressions into the noise power P_N discussed above, we obtain the following expression [17]:

$$P_N = \frac{\hbar\omega\Delta\omega n_{sp}}{2\pi\eta(1 - r_1^2)} [1 - \exp(-(g_m - \alpha_0)L)][1 + r_1^2 \exp((g_m - \alpha_0)L)] \quad (6.117)$$

where the amplifier gain A is given by

$$A = \frac{(1 - r_1^2)(1 - r_2^2)\exp[(g_m - \alpha_0)L]}{|1 - r_1 r_2 \exp(2j\beta_0 L)|^2} \quad (6.118)$$

When the FPA is driven close to the threshold, or for frequencies close to the cavity resonance, we can approximate

$$r_1 r_2 \exp(2j\beta_0 L) \approx 1 \quad (6.119)$$

and hence the noise power is given by

$$P_N = \frac{\hbar\omega\Delta\omega n_{sp}}{2\pi\eta(1 - r_1^2)r_2} [1 - r_1 r_2][r_1 + r_2]A \quad (6.120)$$

The above expressions are identical to those derived in the previous section using the equivalent circuit model.

6.6 SUMMARY

In this chapter the noise induced by random spontaneous emission events in a SLA has been analysed. Three different approaches reported in the literature have been reviewed. It was found that the approach using the photon statistics master equation was the simplest to understand and apply. However, its range of applicability is fairly limited because (i) the analysis becomes more complex as the structure of the SLA becomes more complicated, and (ii) it ignores the effect of the non-uniform material gain profile on the noise characteristics. Nevertheless, this approach allows us to recognise the different components in the total output noise power detected by a photo-detector, namely shot noises due to amplified signal and amplified spontaneous emissions, and the beat noise between the signal and spontaneous emissions. The contribution of shot noise was found to be negligible in most cases. The other two approaches, rate equations and travelling-wave equations, attempted to overcome the short-comings of the photon statistics master equation, but the complexity grew to such a level that no simple analytical non-numerical solutions could be obtained.

On the basis of the equivalent circuit developed in Chapter 5 and by considering the quantum mechanical fluctuations due to the negative resistance in the equivalent circuit, the spontaneous emissions in a SLA can be modelled accurately. This has been verified by analysing the spontaneous emission power emitted from a simple FPA with a uniform gain profile. Instead of using Green's function, identical results can be achieved using the equivalent circuit model with simple circuit and signal analysis. The power of the model has been further illustrated by analysing the effect of stray reflections from the fibre ends on the spontaneous emission power from the FPA. Finally, the structural design of low-noise SLAs and the effect of noise on the performance of a simple optical fibre communication system have been studied. These results may be used in conjunction with the results of Chapters 4 and 5 to devise an optimum structure for a SLA or to optimise the system configuration to achieve an acceptable performance. The equivalent circuit model proved to be extremely useful in this respect because a single model can yield all the necessary results for optimisation (cf. the separation of the analysis of gain and saturation characteristics with that of noise in most literature). A computer-aided design package, for instance, could be built based on this equivalent circuit model together with TMM analysis.

6.7 REFERENCES

[1] D. L. Snyder, *Random Point Processes*, John Wiley and Sons, New York, 1975.

[2] R. Loudon, *The Quantum Theory of Light*, 2nd edition, Oxford University Press, Oxford, UK, 1983.

[3] H. Taub and D. L. Schilling, *Principles of Communication Systems*, 2nd edition, McGraw-Hill, New York, 1986.

[4] N. Wax, *Selected Papers on Noise and Stochastic Process*, Dover, New York, 1954.

[5] K. Kikuchi, C. E. Zah and T. P. Lee, "Measurement and analysis of phase noise generated from semiconductor optical amplifiers", *IEEE J. Quantum Electron.*, **QE-27**, No. 3, pp. 416–422, 1991.

[6] C. H. Henry, "Theory of the phase noise and power spectrum of a single mode injection laser", *IEEE J. Quantum Electron.*, **QE-19**, No. 9, pp. 1391–1397, 1983.

[7] C. H. Henry, "Phase noise in semiconductor lasers", *IEEE J. Lightwave Technol.*, **LT-4**, No. 3, pp. 298–311, 1986.

[8] T. Mukai and Y. Yamamoto, "Noise in an AlGaAs semiconductor laser amplifier", *IEEE J. Quantum Electron.*, **QE-18**, No. 4, pp. 564–575, 1982.

[9] T. Mukai and T. Yamamoto, "Noise characteristics of semiconductor laser amplifiers", *Electron. Lett.*, **17**, No. 1, pp. 31–33, 1981.

[10] Y. Yamamoto, "Noise and error rate performance of semiconductor laser amplifiers in PCM-IM optical transmission systems", *IEEE J. Quantum Electron.*, **QE-16**, No. 10, pp. 1073–1081, 1980.

[11] T. Mukai, Y. Yamamoto and T. Kimura, "Optical amplification by semiconductor lasers", *Semiconductors and Semimetals*, **22**, Part E, pp. 265–319, Academic Press, London, 1985.

[12] G. P. Agrawal and N. K. Dutta, *Long-wavelength Semiconductor Lasers*, Van-Nostrand Reinhold, New York, 1986.

[13] E. L. Goldstein and M. C. Teich, "Noise in resonant optical amplifiers of general resonator configuration", *IEEE J. Quantum Electron.*, **25**, No. 11, pp. 2289–2296, 1989.

[14] G. Bjork and O. Nilsson, "A tool to calculate the linewidth of complicated semiconductor lasers", *IEEE J. Quantum Electron.*, **QE-23**, No. 8, pp. 1303–1313, 1987.

[15] J. A. Armstrong and A. W. Smith, "Intensity fluctuation in GaAs laser emission", *Phys. Rev.*, **140**, No. 1A, pp. A155–A164, 1965.

[16] A. Yarive, *Optical Electronics*, 3rd edition, Holt–Saunders.

[17] C. H. Henry, "Theory of spontaneous emission noise in open resonators and its application to lasers and optical amplifiers", *IEEE J. Lightwave Technol.*, **LT-4**, No. 3, pp. 288–297, 1986.

[18] T. Mukai, Y. Yamamoto and T. Kimura, "S/N and error rate performance in AlGaAs semiconductor laser preamplifier and linear repeater systems", *IEEE Trans. Microwave Theory Technol.*, **MTT-30**, No. 10, pp. 1548–1556, 1982.

[19] H. Kressel, *Semiconductor Devices for Optical Communications*, 2nd edition, Springer-Verlag, Berlin, 1982.

[20] T. Saitoh and T. Mukai, "Recent progress in semiconductor laser amplifiers", *IEEE J. Lightwave Technol.*, **LT-6**, No. 11, pp. 1156–1164, 1988.

[21] J. C. Simon, "GaInAsP semiconductor laser amplifiers for single-mode fiber communications", *IEEE J. Lightwave Technol.*, **LT-5**, No. 9, pp. 1286–1295, 1987.

[22] K. Hinton, "A model for noise processes in semiconductor laser amplifiers, Part 1: the travelling-wave semiconductor laser amplifier", *Opt. Quantum Electron.*, **21**, pp. 533–546, 1989.

[23] K. Hinton, "Optical carrier linewidth broadening in a travelling wave semiconductor laser amplifier", *IEEE J. Quantum Electron.*, **QE-26**, No. 7 pp. 1176–1182, 1990.

[24] K. Hinton, "Model for noise processes in semiconductor laser amplifiers, Part 2: the Fabry–Perot semiconductor laser amplifier", *Opt. Quantum Electron.*, **23**, pp. 755–773, 1991.

[25] T. J. Shepherd and E. Jakeman, "Statistical analysis of an incoherently coupled, steady-state optical amplifier", *J. Opt. Soc. Am. B*, **4**, No. 11, pp. 1860–1869, 1987.

[26] K. Vahala and A. Yariv, "Semiclassical theory of noise in semiconductor lasers—Parts I and II", *IEEE J. Quantum Electron.*, **QE-19**, No. 6, pp. 1096–1109, 1983.

[27] K. Kikuchi, "Proposal and performance analysis of novel optical homodyne receiver having an optical preamplifier for achieving the receiver sensitivity beyond the shot-noise limit", *IEEE Photon. Technol. Lett.*, **4**, No. 2, pp. 195–197, 1992.

[28] P. Spano, S. Piazzola and M. Tamburrini, "Phase noise in semiconductor lasers; a theoretical approach", *IEEE J. Quantum Electron.*, **QE-19**, No. 7, pp. 1195–1199. 1983.

[29] O. Nilsson, Y. Yamamoto and S. Machida, "Internal and external field fluctuations of a laser oscillator: Part II — Electrical circuit theory", *IEEE J. Quantum Electron.*, **QE-22**, No. 10, pp. 2043–2051, 1986.

[30] W. H. Louisell, *Quantum Statistical Properties of Radiation*, John Wiley and Sons, New York, 1973.

[31] M. I. Sargent, M. O. Scully and W. E. Lamb Jr, *Laser Physics*, Addison-Wesley, Wokingham, UK, 1974.

[32] Y. Yamamoto, "AM and FM quantum noise in semiconductor lasers—Part I: Theoretical analysis", *IEEE J. Quantum Electron.*, **QE-19**, No. 1, pp. 34–46, 1983.

[33] C. A. Balanis, *Advanced Engineering Electromagnetics*, John Wiley and Sons, New York, 1989.

[34] K. B. Kahen, "Green's functional approach to resonant cavity analysis", *IEEE J. Quantum Electron.*, **QE-28**, No. 5, pp. 1232–1235, 1992.

[35] H. Ghafouri-Shiraz and C. Y. J. Chu, "Analysis of waveguiding properties of travelling-wave semiconductor laser amplifiers using perturbation techniques", *Fiber and Integrated Optics*, **11**, pp. 51–70, 1992.

[36] H. S. Sommers Jr, "Critical analysis and correction of the rate equation for the injection laser", *IEE Proc.*, **132**, Part J, No. 1, pp. 38–41, 1985.

[37] H. C. Casey and F. Stern, "Concentration-dependent absorption and spontaneous emission of heavily doped GaAs", *J. Appl. Phys.*, **47**, No. 2, pp. 631–643, 1976.

[38] F. Stern, "Calculated spectral dependence of gain in excited GaAs", *J. Appl. Phys.*, **47**, No. 2, pp. 5382–5386, 1976.

[39] D. Marcuse, "Classical derivation of the laser rate equation", *IEEE J. Quantum Electron.*, **QE-19**, No. 8, pp. 1228–1231, 1983.

[40] J. L. Gimlett and N. K. Cheung, "Effects of phase-to-intensity noise conversion by multiple reflections on giga-bit-per-second DFB laser transmission system", *IEEE J. Lightwave Technol.*, **LT-7**, No. 6, pp. 888–895, 1989.

[41]   J. L. Gimlett, M. Z. Iqbal, L. Curtis, N. K. Cheung, A. Fontana and G. Grasso, "Impact of multiple reflection noise in Gbit/s lightwave system with optical fibre amplifiers", *Electron. Lett.*, **25**, No. 20, pp. 1393–1394, 1989.

[42]   L. Gillner, "Comparative study of some travelling-wave semiconductor laser amplifier models", *IEE Proc.*, **139**, Part J, No. 5, pp. 339–347, 1992.

[43]   C. Y. J. Chu and H. Ghafouri-Shiraz, "Effects of stray reflections on performance of semiconductor laser diode amplifiers", *Opt. Laser Technol.*, **26**, No. 6, pp. 403–407, 1994.

[44]   G. Bjork and O. Nilsson, "A new exact and efficient numerical matrix theory of complicated laser structures: properties of asymmetric phase-shifted DFB lasers", *IEEE J. Lightwave Technol.*, **LT-5**, No. 1, pp. 140–146, 1987.

[45]   H. Ghafouri-Shiraz, K. Cameron and R. M. A. Fatah, "Achievement of low coupling loss between a high "NA" MCVD-single-mode fibre by conical microlens", *Microwave Opt. Technol. Lett.*, **3**, No. 6, pp. 214–217, 1990.

[46]   J. John, T. S. M. Maclean, H. Ghafouri-Shiraz and J. Niblett, "Matching of single-mode fibre to laser diode by microlenses at $1.5\,\mu m$ wavelength", *IEE Proc.*, Part J, **141**, No. 3, pp. 178–184, 1994.

[47]   M. Born and E. Wolf, *Principle of Optics*, 6th edition, Pergamon Press, 1983.

[48]   D. Schicketanz and G. Zeidler, "GaAs-double-heterostructure lasers as optical amplifiers", *IEEE J. Quantum Electron.*, **QE-11**, No. 2, pp. 65–69, 1975.

[49]   T. Saitoh and T. Mukai, "$1.5\,\mu m$ GaInAsP travelling-wave semiconductor laser amplifier", *IEEE J. Quantum Electron.*, **QE-23**, No. 6, pp. 1010–1020, 1987.

[50]   T. Li, *Topics in Lightwave Transmission System*, Academic Press, London, 1991.

[51]   J. P. Gordon and L. F. Mollenauer, "Phase noise in photonic communications systems using linear amplifiers", *Opt. Lett.*, **15**, No. 23, pp. 1351–1353, 1990.

[52]   Y. Yamamoto, *Coherence Amplification and Quantum Effects in Semiconductor Lasers*, John Wiley and Sons, New York, 1991.

[53]   P. M. Morse and H. Feshbach, *Methods of Theoretical Physics: Part I*, McGraw-Hill, New York, 1953.

[54]   G. H. B. Thompson, *Physics of Semiconductor Laser Devices*, John Wiley and Sons, New York, 1980.

[55]   A. J. Lowery, "Amplified spontaneous emission in semiconductor laser amplifiers: validity of the transmission-line laser model", *IEE Proc.*, **137**, Part J, No. 4, pp. 241–247, 1990.

[56]   C. Y. J. Chu and H. Ghafouri-Shiraz, "Equivalent circuit theory of spontaneous emission power in semiconductor laser optical amplifiers", *IEEE, J. Lightwave Technol.*, **LT-12**, Vol. 12, No. 5, pp. 760–767, May 1994.

[57]   C. Y. J. Chu and H. Ghafouri-Shiraz, "Analysis of gain and saturation characteristics of a semiconductor laser optical amplifier using transfer matrices", *IEEE J. Lightwave Technol.*, Vol. 12, No. 8, pp. 1378–1386, August 1994.

7

EXPERIMENTAL STUDIES ON SEMICONDUCTOR LASER AMPLIFIERS

7.1 INTRODUCTION

In Chapters 4–6, several characteristics of semiconductor laser amplifiers (SLAs) were studied theoretically. These include: the effective refractive index of the active region and the polarisation dependence on modal gain coefficients (Chapter 4); amplifier gain and output saturation power (Chapter 5); spontaneous emissions from a SLA and their effect on the noise characteristics of the amplifier (Chapter 6). An equivalent circuit consisting of negative resistances and lossless transmission lines has been developed, which can be used as a basis for computer-aided design and analysis of SLAs, even with fairly complex structures. This has been illustrated by using it to analyse the effects of stray reflections from the ends of coupling fibres on the spontaneous emissions generated by SLAs, and on their gain characteristics (see subsection 6.3.3 in Chapter 6). The investigation so far has been purely theoretical. Before this theory can be used for the design of devices or systems, it is necessary to check the validity of its predictions by experimental studies. These experimental studies are arranged in three parts: (i) characterisation of the recombination mechanism in the SLA by measuring its output spontaneous emission power against injection or bias current, (ii) measurement of the gain characteristics; and (iii) measurement of the noise generated by a SLA in the presence of an input signal. These results will be compared with those predicted by theory. The experimental set-up will also demonstrate some of the practical aspects of operating SLAs.

7.2 BASIC SET-UP FOR MEASUREMENTS

Experimental work has been based on a SLA supplied by British Telecom Research Laboratories (BTRL). The centre wavelength of the gain spectrum is

Table 7.1 Structural parameters of the SLA used in the experiment

Length of amplifier	L	=	500 μm
Width of active region	W	=	2.0 μm
Thickness of active region	d	=	0.1 μm
Optical confinement factors	Γ	=	0.16 (TE)
		=	0.13 (TM)
Facet reflection coefficients	$R_1 = R_2$	=	0.01%

1.510 μm and the principal structural parameters are given in Table 7.1. The amplifier is a MOVPE grown buried heterostructure laser coated with multi-layer anti-reflection (AR) coatings, and is designed to be polarisation insensitive. In order to test such a laser amplifier it is necessary to set up a test rig which includes a laser source to provide input signals to the amplifier, the SLA for test, and a photo-detector to measure the light output from the amplifier [1]. The details of mounting these components and the electronics for biasing and for temperature control are described in the following subsections.

7.2.1 *The laser source*

The laser source is a semiconductor distributed feedback (DFB) laser diode, again supplied by BTRL. The emission wavelength is 1.514 μm, and the emission spectrum of this type of laser diode is very stable against small fluctuations in temperature and bias current [2].

As with all types of semiconductor laser diodes, a stable current source is required to bias the device [3]. This current source must be variable to control the output power emitted by the laser. For the initial trial, the circuits were designed to bias the laser diode both in a pulsed regime (this will reduce heating of the p–n junction [4]) and in a continuous wave (CW) regime, which requires a constant current source. A simple circuit can be used to achieve this, which consists of two parts. We first examine the constant current source shown in Figure 7.1. The two n–p–n transistors form a two-stage amplifier which provides the constant current required to bias the laser diode. The two variable resistors VR$_1$ and VR$_2$ are used to adjust the emitter current of the transistor, which will in turn vary the collector current flowing in the transistor. This collector current flows across the laser diode which provides the necessary bias current. Resistors VR$_1$ and VR$_2$ provide coarse and fine control of the laser diode bias current, respectively. The inductor in series with the DFB laser diode provides protection against spurious voltage spikes which may occur during, for instance, turning on the circuit [2]. Also an anti-parallel junction diode is connected across the laser diode to protect it against any accidental reversal of voltage across the laser.

The second part of the circuit is shown in Figure 7.2, which is the modulator circuit used for pulsed operation of the DFB laser diode. The circuit is based on a simple differential amplifier using a long-tail pair with high frequency

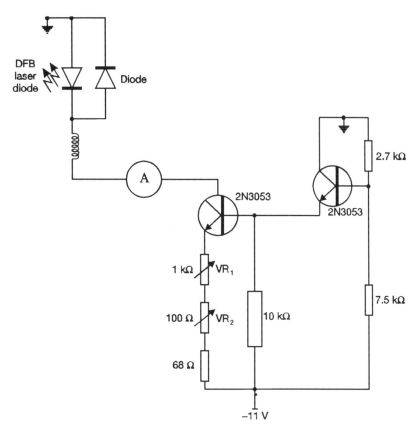

Figure 7.1 Circuit diagram of a current source for bias in both the laser diode and laser amplifier

transistors such as the BFY90 series. For compatibility with the constant current source of Figure 7.1, the output waveform of the modulator should be in ECL levels (i.e. high at -0.8 V and low at -1.6 V) [3–5]. The Zener diodes are used to provide a voltage reference to bias Q_2 and Q_1, and the variable resistor VR is used to adjust the depth of the modulation. In practice, to operate the DFB laser diode under pulsed conditions will require the diode to be pre-biased to a certain level, such that the effect of relaxation oscillation can be minimised [3, 6]. Thus the modulator in Figure 7.2 is connected to the DFB laser diode with the constant current shown in Figure 7.1. Furthermore, most signal generators produce signals with levels between $+5$ V and -5 V. These signals are converted to ECL levels in the modulator by passing first through a LT1016CN comparator to convert to TTL levels and then through a 100124 TTL-ECL converter. For pulsed operation the laser is first biased to its nominal operating current with the controlled current source; the modulator is then adjusted to produce the required depth of modulation. The current source and the modulator are shown in Figure 7.3, where the metal case was used to shield the circuit from external interference.

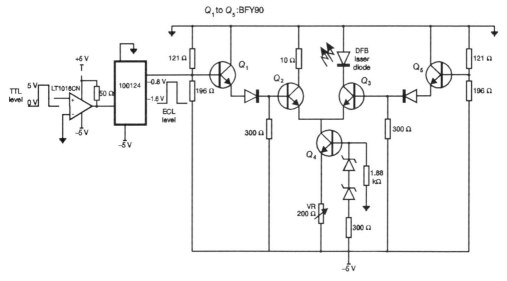

Figure 7.2 Circuit diagram of the modulator used for pulsed operations of the laser diode

Although the emission spectra of DFB laser diodes are more stable than FP laser diodes, their operating temperature has to be stabilised. There are a number of different ways to achieve this goal [3, 7]. One way is to detect the output power of the laser diode by a photo-detector, which is used in a

Figure 7.3 Current source to bias DFB laser diode

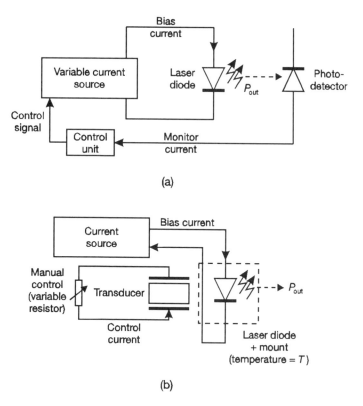

Figure 7.4 Controlling the temperature of the laser diode using: (a) a closed-loop system; (b) an open-loop system

feedback controller to control the bias current of the laser diode. This will eliminate any increase in optical power due to temperature drift, which would subsequently cause an increase in bias current, resulting eventually in *catastrophic breakdown* [2, 6]. However, the feedback has to be calibrated very carefully to maintain the stability of the entire circuit.

A simpler way is to use an open-loop control. A comparison between the open-loop and closed-loop feedback controls is illustrated in Figures 7.4(a) and (b). A sensor that converts heat to electrical current is necessary. A common choice is a Peltier heat pump [8] which is placed between the heat sink and the device that is to be temperature controlled. The heat extracted by the pump is controlled by a bias current (see Figure 7.5). The heat pump is biased at a fixed current. The device under control will be emitting an amount of heat energy Q. According to the magnitude of the bias current, the heat pump will "pump" away excessive heat (by converting it into an electrical current) such that the temperature of the device falls to a designated value T. At equilibrium, therefore, the device can maintain a fixed temperature by controlling the current flowing in the heat pump. The current required to control the temperature for a particular device can be found from a typical set of calibration curves such as the ones shown in Figure 7.5, which are usually

Hot side mounted to heat sink

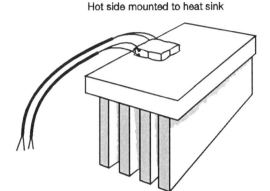

Universal performance graphs

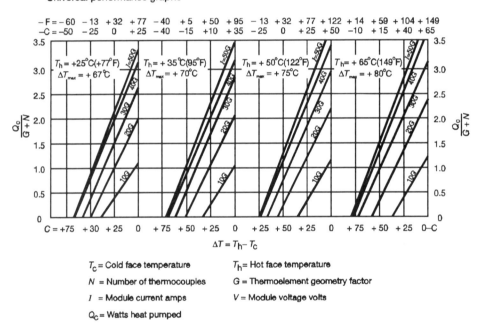

T_c = Cold face temperature T_h= Hot face temperature

N = Number of thermocouples G = Thermoelement geometry factor

I = Module current amps V = Module voltage volts

Q_c= Watts heat pumped

Figure 7.5 A sketch of the Peltier heat pump used as a temperature controller, together with calibration curves

supplied by the manufacturer. The temperature control system shown in Figure 7.4(b), is an open-loop type in the sense that the current flowing into the heat pump has to be controlled manually. However, in practice, whilst closed-loop control is necessary in unmanned optical communication systems, for experimental work an open-loop control is adequate.

To supply current to the heat pump, we can use a similar circuit to that used to bias the laser diode in Figure 7.1. However, although the heat pump requires a very small bias voltage ($V_{max} \leqslant 0.5\,\text{V}$), it needs a fairly large amount of current ($I_{max} \approx 1.2\,\text{A}$). Hence the circuit in Figure 7.1 is not suitable for this

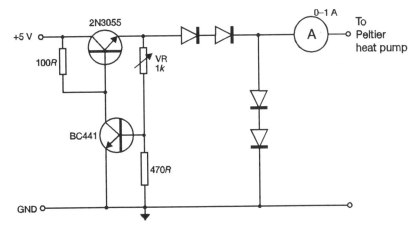

Figure 7.6 Circuit diagram of the temperature controller of the laser and laser amplifier

Figure 7.7   Temperature control for laser diode and amplifier

purpose. Instead, the circuit shown in Figure 7.6 (see also Figure 7.7) was used where the VR controls the pump current and is adjusted manually to maintain the desired temperature at the laser. Figure 7.8 shows how the laser diode is mounted on a micro-positioner. The laser is fixed to a small copper plate attached to the Peltier heat pump, which is isolated thermally by a perspex block so that it is only the excess heat from the laser that is pumped away. The

Figure 7.8 Mount for the DFB laser diode

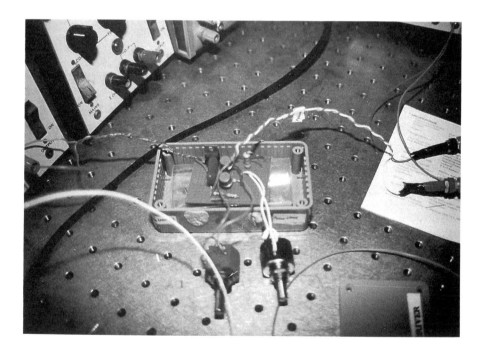

Figure 7.9 Constant current source used to bias the laser amplifier

Figure 7.10   Mount used for the semiconductor laser amplifier

mount is sprayed black to minimise stray reflections, as such reflections can affect the stability of the emission spectrum [9].

7.2.2 *Semiconductor laser amplifier*

The SLA used in the tests has to be biased with a constant current and also has to be stabilised in temperature. The controlled current source is the same as in Figure 7.1 and the Peltier heat pump circuit is as in Figure 7.4. The actual circuit constructed for the current source is as shown in Figure 7.9. The mounting is similar to the laser diode (see Figure 7.10) differing only in that the SLA is raised to facilitate the coupling of the input and output signals.

7.2.3 *Detection circuit*

Two types of detector were required for the experimental study. A precision optical power meter, by Anritsu, with its own optical detector head was used for most of the measurements. This detector head has a relatively large surface area, which improves its sensitivity and coupling efficiency. For noise measurements it is necessary to use a simple detection circuit consisting of a photo-detector connected to a load resistor [3]. The photo-detector is an InGaAsP *pin* photodiode supplied by British Telecom Research Laboratories. This device has a quantum efficiency approaching 95%. One advantage in

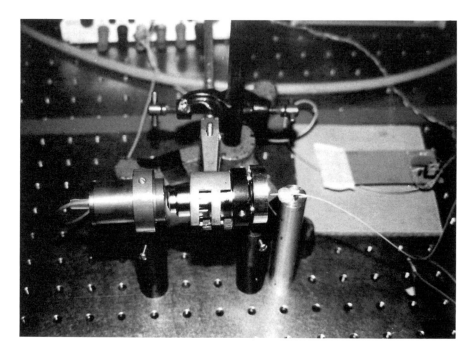

Figure 7.11   Detector set-up using the pin photodiode supplied by BTRL, showing also the simple detection circuit at the back

using this type of photodiode is that the shot noise due to thermal fluctuations, and the dark current in the load resistor, are low [6, 7]. However, the surface area of this photodiode is quite small (the diameter of the active area is approximately 50 μm), therefore care has to be taken to obtain good coupling efficiency between the signal and the photodiode. This photodiode is mounted on a copper cylinder, and connected as shown in Figure 7.11. Two additional optical components are included in this assembly; a narrow band optical filter with a bandwidth of around 10 nm and a lens. The filter is used to minimise unwanted spontaneous emission noise generated by the amplifier, whilst the lens is used to improve the coupling efficiency between the output from the filter to the photo-detector.

7.3   EXPERIMENTAL STUDIES ON RECOMBINATION MECHANISMS

The objective of the first set of experiments is to investigate the recombination mechanisms in the SLA. This is necessary because we saw in Chapter 3 that there are two possible models for the carrier recombination rate R. The bi-molecular model which takes into account radiative recombination due to spontaneous emissions only: if the active region is undoped,

$$R = Bn^2 \tag{7.1}$$

The other possible model will include the effects of Auger recombination and recombination with defects and impurities near the surface of the material as well [2, 10] (see also the discussion in subsection 3.2.2 in Chapter 3) for which

$$R = A_{nr} n + Bn^2 + Cn^3 \tag{7.2}$$

where A_{nr}, B and C are the recombination coefficients for non-radiative recombination of carriers with defects and impurities, radiative recombination owing to spontaneous emissions, and non-radiative recombination of carriers owing to Auger processes, respectively. In the above equations, n is the injected carrier density, and we have assumed that the active region of the SLA is undoped. Before further analysis of the device under study can be made, it is necessary to determine which of these two models should be used. In this section an experimental determination of the recombination mechanisms will be described such that we can determine whether we should use equation (7.1) or (7.2) in subsequent analyses.

7.3.1 Principles of the experimental measurement

As discussed in the previous chapter (subsection 6.3.2), if a SLA is biased by a finite injection current in the absence of an input signal, a finite amount of optical power will be detected at the output owing to amplified spontaneous emissions (ASEs). This is because the random spontaneous emissions generated by the amplifier are amplified by the optical gain provided by the injection current. The amount of spontaneous emission power detected will be proportional to the optical gain in the active region. On the other hand, the magnitude of the optical gain will also depend on the injection current. We have seen in Chapter 5, subsection 5.3.2, that at a distance z within the active region, the modal gain coefficient $g(z)$ is related to the injection current density j via the rate equation [11]

$$\frac{j}{ed} = R(z) + \frac{c\Gamma}{N_g \Delta E} \int_0^{\Delta E} g(z, E)\, S(z, E)\, \mathrm{d}E \tag{7.3}$$

where d is the active layer thickness, E is the energy of the photon, Γ is the optical confinement factor, N_g is the group effective index of the active region, S is the photon density at z, and $\Delta E = h\Delta f$, where Δf is the optical bandwidth of the material. The carrier recombination rate R can take the form of either equation (7.1) or equation (7.2), depending whether Auger recombination is important. Therefore equation (7.3) can be solved in terms of the carrier density n for a particular injection current density j, in which the modal gain coefficient can be found from the material gain coefficient g_m:

$$g(z, E) = \Gamma[g_m(z, E) - \alpha_a(z, E)] - (1 - \Gamma)\alpha_c(z, E) \tag{7.4}$$

where we have assumed that the optical confinement factor Γ can be well defined ([12]; see also Chapter 4). This is true in the present case as the SLA supplied by BTRL is a buried heterostructure. The solution procedures of these equations have already been outlined in subsection 5.3.2 and Section 5.4. The peak-gain material gain coefficient for a particular value of n can be found either by the well-known linear model

$$g_{\mathrm{m}} = A(n - n_0) \qquad (7.5)$$

or by the parabolic model suggested by Ghafouri-Shiraz [13] (subsection 3.2.2 in Chapter 3):

$$g_{\mathrm{m}} = an^2 + bn + c \qquad (7.6)$$

Knowing n for a particular injection current and hence the peak material gain g_{m} (from either equation (7.5) or (7.6)) and the peak modal gain coefficient g allows us to calculate the output spontaneous emission power P_{out} generated by the amplifier at the peak wavelength of the material gain spectrum (known hereafter as the *peak-gain wavelength*). Obviously the amount of P_{out} generated by a SLA biased at particular injection current density j will then be determined by the model used for the recombination rate R, as discussed by Wang *et al.* [14]. Hence by measuring the actual value of P_{out} at the peak-gain wavelength for the device with parameters listed in Table 7.1 [15, 16], and comparing the measurements with the theoretical values predicted by the two recombination models, we can then see which model, equation (7.1) or (7.2), is the appropriate one to be used in subsequent analysis [31].

7.3.2 Experimental procedures

The laser amplifier with parameters tabulated in Table 7.1 was mounted and connected to the current source and the temperature controller. The temperature was maintained at room temperature (20°C) by biasing the Peltier heat pump at a fixed current (around 400 mA), which can be determined from the calibration curves in Figure 7.5. A fixed injection current was used to bias the amplifier. The peak-gain wavelength of the amplifier supplied by the manufacturer is 1.510 μm, and the output power of the amplifier was measured at this wavelength by the Anritsu optical power meter. To minimise coupling loss between the amplifier output and the detection head of the power meter, the detector had to be as close as possible to the amplifier *without* hitting the facet and damaging the device. This was done by aligning it with a micro-positioner and the aid of a microscope. The output spontaneous emission power is measured by varying the injection current from 10 mA to 80 mA in 5 mA steps.

7.3.3 Results and discussions

The measured results are shown as the solid circles in Figure 7.12 [31]. They are compared with the theoretical spontaneous emission power outputs for the two recombination models given by equations (7.1) and (7.2). The theoretical

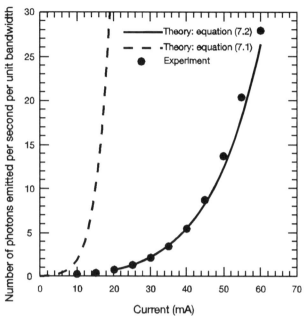

Figure 7.12 Measured and calculated number of spontaneous emission photons generated by the SLA at different bias current levels. Two recombination models have been used for comparison

results calculated by considering radiative recombination owing to spontaneous emissions only are shown as the solid line in Figure 7.12, whereas the results calculated by including Auger recombination and recombination with defects are shown as the dashed curve. They are calculated by using the averaged photon density (AVPF) approximation proposed by Adams [11] with $\beta = 10^{-4}$. The additional parameters required for the calculations are tabulated in Table 7.2. Sources for some of these parameters are shown in the table. The experimental results agree well with those predicted by the recombination rate described by equation (7.2), i.e. with non-radiative recombination included. It can be seen that if the carrier recombination is due to radiative spontaneous emissions only, then the theoretical results will over-estimate the amount of spontaneous emission generated by the amplifier for each particular value of

Table 7.2   Material parameters used in the analysis of the SLA

Differential gain coefficient	A	=	4.63×10^{-16} cm^2
Transparency carrier density	n_0	=	8.31×10^{17} cm^{-3}
Recombination coefficients	A_{nr}	=	10^{-8} s
	B	=	10^{-10} cm^3/s
	C	=	7.5×10^{-29} cm^6/s
Loss coefficient in active region	α_a	=	25 cm^{-1}
Loss coefficient in claddings	α_c	=	50 cm^{-1}

injection current. In fact, this model will predict an onset of oscillation when the amplifier is biased at around 10 mA, which does not happen at all in this experiment. On the other hand, the full recombination model taking into account Auger recombination and recombination with defects gives a much more accurate prediction over the range of measurement (20–60 mA), therefore we will use equation (7.2) for the carrier recombination rate R.

A note about the peak-gain coefficient model. Two models are possible, which have been discussed in equations (7.5) (the linear model) and (7.6) (the parabolic model) already. Both have been used in analysing the above measurements. However, there are no significant differences between using either model. For simplicity, the theoretical results plotted on Figure 7.12 are calculated using the simple linear model of equation (7.5), and this model will be used hereafter for other analyses. This does not imply that there is no difference between the two models. As seen from Figure 3.2, subsection 3.2.2 in Chapter 3, there is a significant difference between the values of the peak-gain coefficient calculated by these two models when the injection carrier density n is close to the transparency carrier density n_0. However, because of the losses in the active region and claddings in actual devices, the value of n required to create a significant value of optical gain in the active region will be well in the region of n where both models do not differ significantly.

7.4 MEASUREMENT OF GAIN CHARACTERISTICS

7.4.1 Experimental set-up

The next set of measurements is to determine the gain of the amplifier for different input signal levels and different magnitudes of injection current. The DFB laser source used has to be set up as discussed in subsection 7.2.1. A measurement of the output power of the DFB laser diode against its bias current has first been taken, which is plotted on Figure 7.13. It can be seen that the stability of the DFB laser diode is extremely good, with no "kinks" observed on the light power–current curve (i.e. no mode jumping) [3, 7]. The threshold current of the laser diode is around 15 mA at room temperature. By varying the bias current to the DFB laser diode, its output power can be changed as described by the curve in Figure 7.13. This is used as the variable input optical signal source to the amplifier for the following measurements.

The experimental set-up for measuring the amplifier gain is shown in Figure 7.14 [1, 8]. The output of the DFB laser diode, which is used as an input signal source for the amplifier, is coupled to an isolator. This isolator, shown in Figure 7.15, is a polarisation beam-splitting cube followed by a quarter-wave plate. The principle of this isolator is illustrated in Figure 7.16. Any reflected signals towards the DFB laser will be converted, after the quarter-wave plate, to signals with a polarisation orthogonal to that of the input signal to the isolator. The polarisation beam-splitting cube will ensure that this signal will not be able to return to the laser diode, thereby isolating the emitted beam

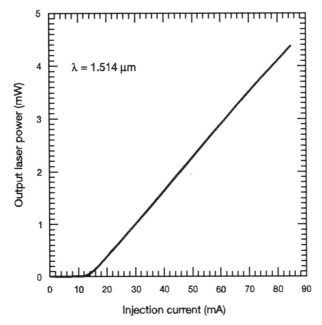

Figure 7.13 Measured output DFB laser diode output power versus bias current

from reflected beam. Because the signal emerging after the isolator is circularly polarised [17], it has to be converted back to a linearly polarised signal before coupling into the SLA. This can be achieved by an additional quarter-wave plate, as shown in Figures 7.14 and 7.15. By rotating this quarter-wave plate we can control the plane of polarisation of the linearly polarised light emerging from the plate [17]. This will be used to control the state of polarisation of the input signal to the amplifier when the polarisation dependence of the amplifier

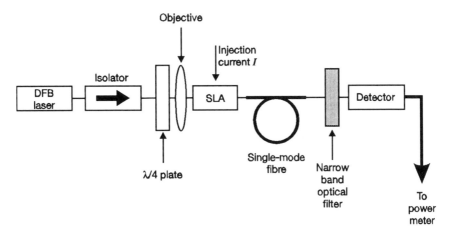

Figure 7.14 Experimental set-up for measuring gain and output saturation power in a SLA

Figure 7.15   Isolator and retardation plates used to couple signal to the amplifier

gain is to be measured [18]. An objective is used to couple the linearly polarised light to the laser amplifier, as shown in Figure 7.15.

The output of the amplifier is butt coupled to a piece of single-mode fibre having a length of 2 m (Figure 7.17). This fibre was supplied by BTRL and was designed specially with a high numerical aperture [7, 19] to improve the butt coupling efficiency between the amplifier and the optical fibre [20, 21]. The loss of the optical fibre is 0.25 dB/km, according to the manufacturer. The coupling efficiency between the output signal and the detector head can be improved by the insertion of this length of low loss optical fibre because the matching of the spot sizes can be improved. Finally, before detection, the signals are passed through a narrow band optical filter of bandwidth 10 nm centred at $1.50\,\mu$m to eliminate excessive spontaneous emission power which would otherwise affect the measurements. Alternatively, a lock-in amplifier can be used for this purpose, but its sensitivity is not as good as the Anritsu power meter. Therefore the latter equipment has been used. This is shown in Figure 7.18. The entire set-up is displayed in Figures 7.19 and 7.20.

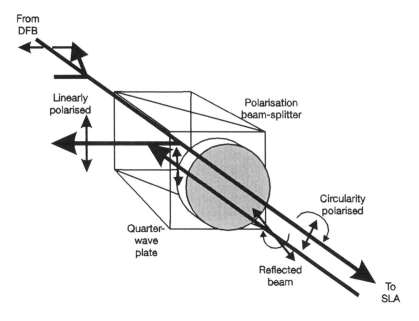

Figure 7.16   Principle of the isolator used in the experimental measurements

Figure 7.17   Fibre coupling to the output of the amplifier

Figure 7.18 The signals are passed through a narrow band optical filter (10 nm bandwidth centred at 1.5 μm wavelength)

Figure 7.19 The complete experimental set-up used to measure amplifier characteristics

Figure 7.20 A close-up of the laser source–laser amplifier used in the experimental set-up

7.4.2 *Experimental procedures*

Determination of coupling losses in the set-up

The first part of this set of measurements will involve determination of the coupling losses within the set-up shown in Figure 7.14, which will be required for the measurement of gain in the following measurements. We will assume that the coupling losses are independent of the signal power such that one set of measurements based on a particular optical power level will be sufficient. This measurement has to be performed simultaneously when the alignment and coupling of the devices and optical components is undertaken. The amplifier is connected first. The SLA is biased at 40 mA at a temperature of 20°C (i.e. room temperature), which generates spontaneous emission power at the output. The power emitted immediately after the facet of the SLA has been measured in subsection 7.3.3, and is plotted in Figure 7.12. The optical components can then be added one by one: first the retardation plate/objective combination, secondly the isolator, thirdly the fibre, and then the optical filter/detector head combination. When each component is added, alignment with the three-axes micro-positioners is made to ensure maximum coupling (in the case of coupling between the fibre and the amplifier output, additional help from a microscope may be necessary). This is done by measuring the optical power after adding each optical component and adjusting the alignment until a maximum optical power is recorded. This measured maximum power will also be used to

Table 7.3   Results of coupling efficiencies measurements

Coupling efficiency between DFB laser and isolator	η_1 =	−6.17 dB
Coupling efficiency between isolator and quarter-wave plate/lens combination	η_2 =	−3.75 dB
Coupling efficiency between lens and SLA	η_{c1} =	−7.64 dB
Coupling efficiency between fibre and SLA	η_{c2} =	−11.37 dB
Coupling efficiency between fibre and filter/detector	η_3 =	−14.63 dB

N.B. The losses in the low-loss single-mode fibre been neglected.

calculate the coupling loss between the optical components. If the previous measured maximum optical power before addition of the extra optical component is P_2, then the coupling loss η_c between these optical components is simply given by:

$$\eta_c = \frac{P_2}{P_1} \qquad (7.7)$$

In this manner, the coupling loss between different optical components in the experimental set-up shown in Figure 7.14 can be determined. The measured coupling losses are tabulated in Table 7.3. These values will be used in the following experimental measurements.

Measurement of amplifier gain

The output of the SLA detected by the optical power meter will consist of both the power of the spontaneous emissions that pass through the optical filter as well as the signal. Therefore before any measurement is taken, the amplifier is set up without any input signal, and the corresponding values of the spontaneous emissions reaching the optical power meter are noted. To measure the amplifier gain, the following procedures are followed. First, the DFB laser diode is biased by a particular current, with its temperature maintained at room temperature. Similarly, the SLA will also be biased at a fixed current. The coupling of the DFB laser diode with the SLA is maximised by proper alignment with the three-axes micro-positioner (Figure 7.20). The power P emitted by the DFB laser diode for a given current can be found from the curve in Figure 7.13. The true input power P_{in} to the optical amplifier will be $\eta_1\eta_2\eta_{c1}P$, where η_1 is the coupling efficiency between the isolator and the DFB laser, η_{c1} is the coupling efficiency between the quarter-wave plate/ objective with the facet of the laser amplifier. The temperature of the amplifier is then adjusted until a maximum power is detected by the detector head of the optical power meter. This is the case when the amplifier's peak-gain wavelength is matched with that from the DFB laser and the centre wavelength of the optical filter. The quarter-wave plate/objective combination (Figure 7.15) is then rotated. Rotating the quarter-wave plate/objective combination will *not* affect P_{in}, only the state of polarisation of the input signal. This will affect the amplifier gain G as discussed in Chapter 4. The value of G is a maximum when the state of polarisation of the input signal is TE, whereas when the input signal is TM polarised, G will become a minimum. Thus the state of polarisation of the input signal to the SLA can be adjusted by rotating the quarter-wave plate/objective combination until the detected power is a maximum (i.e. TE mode) or minimum (i.e. TM mode). If the detected power of the signal is P_D (i.e. after the spontaneous emission power has been deducted), the true output power of the SLA P_{out} is given by $P_D/(\eta_{c2}\,\eta_3)$ assuming that the loss in the optical fibre is negligible, where η_{c2} is the coupling efficiency between the SLA and single-mode fibre, and η_3 is the coupling efficiency between the optical fibre and the narrow band filter. The values of all of these coupling efficiencies have already been tabulated in Table 7.3. Then the amplifier gain G is simply given by the ratio P_{out}/P_{in}. The resulting values of P_D are measured by the optical power meter. The bias current to the DFB laser diode is varied such that the output power of the amplifier is incremented from $-40\,\text{dBm}$ in steps of $-5\,\text{dBm}$ until a maximum value, which corresponds to the maximum power input to the amplifier. By noting the currents required to bias the DFB laser diode to achieve these output optical powers, P can be found from Figure 7.13. Using the arguments above, the values of P_{out} and P_{in} can be determined, and hence G can be found for various output power levels. The measurements are repeated for different values of the bias current to the SLA ranging from 30 mA to 80 mA in 5 mA steps. Hence the variation of G with the injection current i to the SLA for a particular input power level can also be measured.

7.4.3 Results and discussions

The measured values of the TE mode gain G versus P_{out} are plotted in Figure 7.21 for $i = 20, 30, 40, 50$ and 60 mA, and similar results are plotted in Figure 7.22 for the TM mode. The theoretical results (solid curves) are calculated from the equivalent circuit model using transfer matrix analysis. This theoretical model, as discussed Chapters 5 and 6, will be able to take account of the non-uniform material gain profile in the SLA because of the increasing photon density with z along the amplifier [22]. To assess the accuracy of this model, the amplifier gain G is also calculated using the AVPD approximation proposed by Adams [11] as discussed in subsection 5.3.2 in Chapter 5. The results calculated by this method are plotted as the dashed lines in Figure 7.23 for $i = 40$ mA. The structural parameters used in the analysis have already been tabulated in Table 7.1. In the theoretical analyses, the optical confinement factors shown in Table 7.1 are obtained from the curves shown in Figure 4.10(b) in Chapter 4. The confinement factors for the slab TE and TM modes can be used here because the dimensions of the active region, also shown in Table 7.1, satisfy the single-transverse mode condition (i.e. the field is cut off along the x-direction) [23]. As discussed in Section 7.3 above, the carrier recombination rate used in these analyses will include Auger recombination and recombination with defects. Moreover, in the calculations we found that the

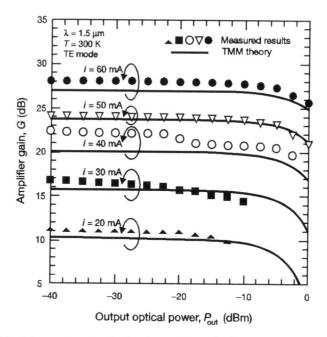

Figure 7.21 Measured and calculated amplifier gain for the TE mode versus output optical power from the SLA with bias current as a parameter. The solid curves represent results calculated by the transfer matrix method (TMM)

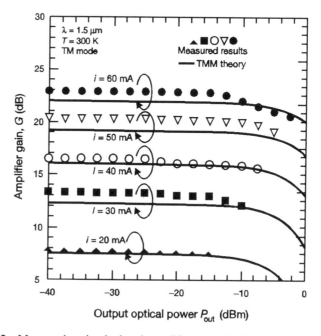

Figure 7.22 Measured and calculated amplifier gain for the TM mode versus output optical power from the SLA with bias current as a parameter. The solid curves represent results calculated by the transfer matrix method (TMM)

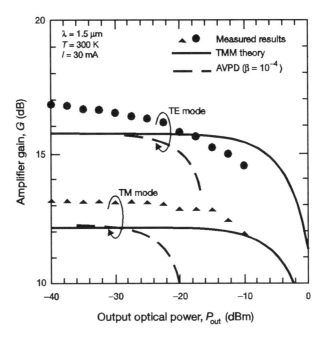

Figure 7.23 Comparison between amplifier gain G versus its output optical power calculated by the TMM and the average photon density (AVPD) approximation

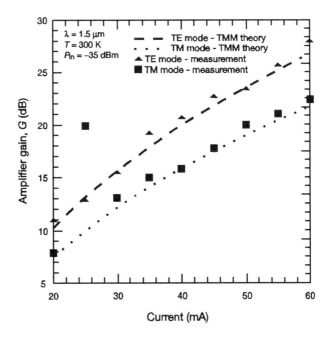

Figure 7.24 Variation of the amplifier gain versus bias current

theoretical results agree better with the measured ones if the dependence of the facet reflectivity on the state of polarisation of the signal is taken into account [24, 25].

It can be seen from these figures that the measured results agree reasonably with those predicted by the TMM theory. A reduction in amplifier gain G is observed as P_{out} increases. This saturation phenomenon has been predicted qualitatively in Chapter 2 using a simple two-level system, which also explains the saturation observed in Figures 7.21 and 7.22. The output saturation power measured for different values of i are plotted in Figure 7.24, which also shows the predicted values obtained from the equivalent circuit model and transfer matrix analysis proposed by the author [32].

The differences between the amplifier gain G for the TE and TM modes are not significant. The polarisation sensitivity of the amplifier, which is defined as the difference between the unsaturated value of G for the TE and TM modes, respectively, is observed to be within the range of 3.4 dB from Figures 7.21 and 7.22. This agrees well with the data supplied by BTRL. However, as seen from Figures 7.21 and 7.22, the saturation behaviour of these two polarisations is quite different. It can be seen that for the TM mode saturation occurs earlier because the optical confinement factor for this mode is smaller [26].

It is interesting to compare the experimental measurements with the theoretical results predicted by the TMM and the AVPD approximation in Figure 7.23. It can be seen that the AVPD gives an unsaturated gain which is close to that predicted by the TMM. However, the AVPD predicts saturation output powers which are much smaller than those obtained from the measurements and from the theoretical results predicted by the TMM. This can be due to the fact that the facet reflectivity of the amplifier under test is relatively low, and the photon density distribution inside the amplifier becomes increasingly non-uniform. The AVPD approximation will not work well under such circumstances, whereas the TMM, by taking care of this variation, can be more accurate. This supports the use of the more accurate analysis using the TMM proposed by the present author.

7.5 MEASUREMENT OF NOISE CHARACTERISTICS

7.5.1 Experimental set-up

This is the most delicate set of measurements to be made. There are two reasons for this; first, the alignment between the optical components and devices has to be optimised carefully stage by stage to obtain maximum coupling efficiency, as with the previous set of measurements. In the present measurement of noise properties we have to measure the r.m.s. values of the fluctuations in the detector current flowing across the load resistor in the photo-detector circuit. Thus additional electronic and optical components will be involved, as discussed in subsection 7.2.3 (Figure 7.11).

The experimental set-up for noise measurements is shown in Figure 7.25. This set-up was used by Mukai and Yamamoto [27, 28] in their noise

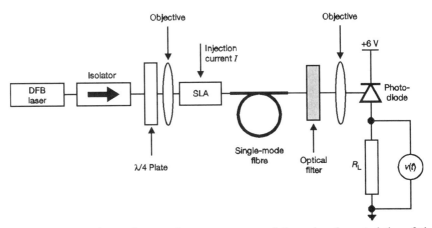

Figure 7.25 Experimental set-up for measurement of the noise characteristics of the SLA

measurements, which was modified from the set-up used to measure intensity fluctuations of a GaAs diode laser by Armstrong and Smith [29]. An optical signal has to be coupled into the SLA, and both the bias current to the SLA and the input optical power (i.e. the output power from the DFB laser source) has to be varied. Thus most of the set-up is identical to that used in gain measurements. The major difference is in the detection scheme. In the present case, the output of the SLA will pass through the single-mode optical fibre and the optical filter, which is then incident on the InGaAsP pin photo-diode. An objective is placed between the photodiode and the output of the filter to improve coupling. The resulting current generated by the detected optical power, which includes both the amplified signals and the spontaneous emissions generated from the SLA, will pass through the load resistor R_L. The fluctuations in the detected power will then result in fluctuations in the voltage across the load resistor, and this can be measured by a high sensitivity r.m.s. voltmeter.

7.5.2   *Experimental procedures*

Because we have changed from the detector head to a pin photodiode, the coupling efficiency between the output of the fibre and the photodiode has to be re-measured. This follows the same procedures as in the previous measurement. The SLA is biased with no input signal, and the power of the spontaneous emissions after the fibre and before the photodiode are measured. The coupling efficiency depends strongly on the alignment between the photo-diode and the fibre. They have to be aligned both horizontally and vertically (to focus the beam from the objective onto the active surface of the photodiode). The maximum coupling efficiency between the fibre and the photodiode is achieved when the measured spontaneous emission power after the objective is maximised. This was measured to be $-27.68\,$dB. Compared with the previous set-up, there is more coupling loss in the present set-up because there are more optical components. This introduces more mis-alignment into the set-up. Although such mis-alignments can be minimised by proper alignment with the micro-positioners and optical mounts, it cannot be totally eliminated. The more optical components the set-up possesses, the more difficult it is to eliminate mis-alignment. Furthermore, there are finite losses (not owing to coupling but because of, for instance, scattering and absorption) in the optical components. An increase in the number of optical components will definitely increase the total loss in the system.

As before, the DFB laser diode is biased at different currents, with its temperature stabilised at room temperature to give a variable optical output power. The SLA is biased at a constant level. As with the previous set of measurements, the quarter-wave plate/objective combination before the input to the SLA is used to adjust the state of polarisation of the input light signal to the amplifier. The temperature of the SLA will be adjusted such that its peak-gain wavelength will coincide with that of the DFB laser source and that of the centre wavelength of the optical filter. The bias current to the DFB laser source will then be varied as in the previous gain measurements to give an output optical power from $-40\,\mathrm{dBm}$ to about $0\,\mathrm{dBm}$ in steps of $5\,\mathrm{dBm}$. The corresponding input power to the SLA can be found from the curve in Figure 7.13 and the coupling efficiency data tabulated in Table 7.3. The r.m.s. voltage across the load resistor R_L is then measured for each of these input power levels. The measurement is then repeated for different SLA bias current levels, ranging again from $20\,\mathrm{mA}$ to $80\,\mathrm{mA}$ in steps of $10\,\mathrm{mA}$. The mean square voltage recorded across the load resistor R_L in an *electrical* bandwidth of $\Delta\omega$ will be given by [26, 28]

$$\langle v^2(t)\rangle = [\eta_{c2}^2\,\eta_3^2\,\eta_\mathrm{D}^2\,\langle i_\mathrm{beat}^2\rangle + 2e\eta_{c2}\,\eta_3\,\eta_\mathrm{D}\langle i_\mathrm{pho}\rangle$$
$$+ 2e\eta_\mathrm{D}\langle i_\mathrm{dark}\rangle]\cdot R_\mathrm{L}\cdot\Delta\omega + 4kT\Delta\omega/R_\mathrm{L} \tag{7.8}$$

In the above equation, $\langle i_\mathrm{beat}^2\rangle$ is the beat noise power per unit bandwidth generated in the photodiode with $\eta_\mathrm{D}=1$ by the beating between the signal and the spontaneous emission components and between the spontaneous emission components *after* filtering by the optical filter with bandwidth $\Delta\omega$. $\langle i_\mathrm{pho}\rangle$ is the steady-state photo-current generated by the photodiode after detecting the output from the SLA. $\langle i_\mathrm{dark}\rangle$ is the dark current of the photodiode if its quantum efficiency is unity [6]. η_D is the actual quantum efficiency of the photo-diode, and the product $\eta_{c2}\,\eta_3$ is the total coupling efficiency between the SLA output and the photodiode. R_L is the load resistance of the photo-detector [30]. Thus the four terms in the above equation represent the voltage fluctuation induced by (i) the beat noise current generated by the SLA output; (ii) the shot noise current generated by the SLA output; (iii) the shot noise current generated by the dark current of the photodiode; and (iv) the shot noise current generated by the thermal noise of the resistive part of the detection circuit, respectively. It can be seen that the choice of the value of the load resistor R_L is vital in the measurement. If the selected value is too small, then the r.m.s voltage becomes too small to measure accurately, but the noise power owing to the dark current and thermal noise from the resistive part of the circuit will also be very small. The sensitivity of the measurement can be improved by increasing the value of R_L, but the noise power due to the dark current and thermal noise becomes very large. However, we can measure the noise power

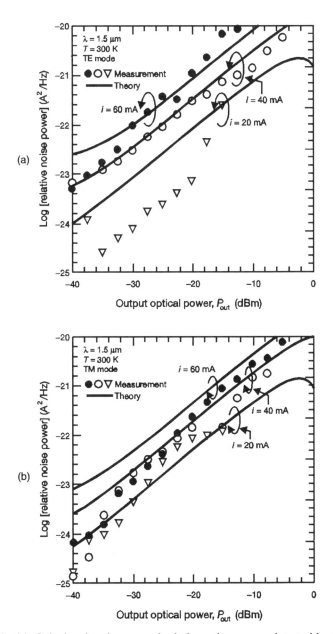

Figure 7.26   (a) Calculated and measured relative noise powers detected by the photo-detector for the TE mode. (b) Calculated (solid curves) and measured relative noise power for the TM signal mode

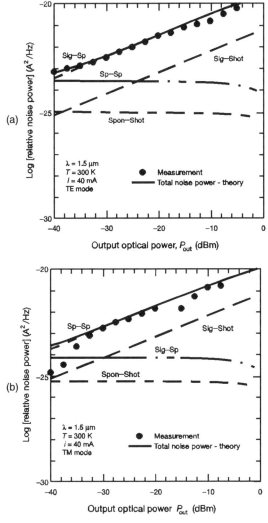

Figure 7.27 (a) Calculated (solid curves) and measured relative noise power for the TE signal mode. Also shown are different components contributing to the total noise power, that is:

(i) signal to spontaneous emission beat noise;
(ii) spontaneous to spontaneous noise due to beating between different spontaneous emission components;
(iii) signal shot noise;
(iv) shot noise due to spontaneous emissions.

(b) Calculated (solid curves) and measured relative noise power for the TM signal mode. Also shown are different components contributing to the total noise power, that is:

(i) signal to spontaneous emission beat noise;
(ii) spontaneous to spontaneous noise due to beating between different spontaneous emission components;
(iii) signal shot noise;
(iv) shot noise due to spontaneous emissions

owing to the shot noise generated by the dark current and thermal fluctuations by measuring the r.m.s. voltage across the load resistor with the photodiode shielded from any stray light. Therefore, for sensitivity reasons (remember that there is quite a significant loss between the output of the SLA and the detector), a relatively large resistance of $R_L = 100 \, \text{k}\Omega$ has been used in the measurements, with the resulting noise power induced by the dark current and thermal fluctuations measured before the output from the SLA is coupled to the photo-diode. The noise power which is due to the SLA can be obtained by subtracting the mean-square voltage across the load resistor R_L, without a signal incident on the photodiode from the total mean-square voltage when the signal is present.

7.5.3 *Results and discussions*

The measured noise power per unit electrical bandwidth has been plotted in Figure 7.26 (a) and (b) versus the input optical signal power for different bias currents, for both the TE and TM modes. This quantity has been estimated from the measured voltage fluctuations across the load resistor by considering its frequency components (i) that lie within the bandwidth of the r.m.s. voltmeter. This has been done under two assumptions: that the bandwidth of the r.m.s. voltmeter is smaller than that of the detection circuit (according to the manufacturer, the bandwidth of the r.m.s. voltmeter is approximately 1 MHz, whereas for the load resistance we have been using the bandwidth of

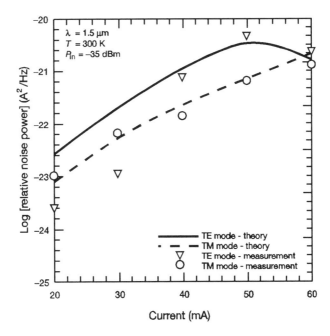

Figure 7.28 Calculated and measured relative noise power of SLA versus bias current for both the TE and TM modes

the detection circuit is estimated to be 1 GHz if the capacitance of the photo-diode is similar to other pin photo-diodes of around 1 pF [3]); and (ii) the frequency responses for both the r.m.s. voltmeter and the detection circuit can be assumed to be that of simple first-order systems with ideal low-pass characteristics. Under these assumptions, only the power of those frequency components of the total fluctuations of the voltage across the load resistor that fall within the pass band of the frequency response of the r.m.s. voltmeter will be measured. The power of the frequency components of the voltage fluctuations outside the pass band of the r.m.s. voltmeter is assumed to be negligible as they have been cut off significantly by the low-pass filter characteristics. In general, total noise power increases with output optical power, as observed from experimental measurements.

The theoretical results of the noise power per unit electrical bandwidth have also been calculated using analytical expressions (equation (6.22)), section 6.2.1 in Chapter 6, derived from the photon statistics master equation. It should be noted that in these calculations both the gain of the amplifier and the single-pass gain used in the calculation by equation (6.22) have been obtained from the field profile generated by the TMM (see Figure 5.16(b) in Chapter 5), instead of using the analytical expressions for the amplifier gain and assuming that the single-pass gain G_s is simply given by $\exp(gL)$, where g is the modal gain coefficient. This is intended to improve the accuracy of the theoretical results. As can be seen from Figure 7.26, the measured results do not agree with the theoretical results as well as those obtained in the gain measurements in the previous section. This may be because of the fact that the estimation of the frequency response of the r.m.s. voltmeter may not be very accurate. However, with these rather crude assumptions, the measured results still illustrate the correct trend in the behaviour of noise, as well as giving the right order of magnitude (remember that we are plotting with logarithmic scales). Furthermore, the measured results are not poor for all currents. It can be seen from Figure 7.27(a) and (b) that the agreement is quite good for $i = 40$ mA. Also shown is a breakdown of the theoretical results into four different components, clearly illustrating the increasing importance of the beat noise between the signal and the spontaneous emissions. It should also be noted that because the polarisation sensitivity of the SLA under study is not very high, the differences in noise performances for both the TE and TM modes for this amplifier are not significant, as supported both by the experimental results (in which some overlapping occurs) and the theoretical predictions.

The noise power per unit electrical bandwidth has also been plotted versus the injection current of the SLA in Figure 7.28 for both the TE and TM modes. The noise power initially increases with i, which can be observed from both the theoretical and experimental results. This can be explained by the increase in amplifier gain and hence the enhancement of the beating between the signal and spontaneous emissions (the major noise contribution) as i increases. There is some discrepancy between the measured results and the theoretical results for the TE measurements, but the general trend can be observed. It can be seen that, as observed in many previous experiments, the noise power is quenched when the injection current increases (notice the hump in the theoretical results for the TE mode). This can be explained by gain saturation, which will suppress the beating between the signal and spontaneous emissions, thereby reducing the overall noise power.

7.6 SUMMARY

The recombination mechanisms, the gain characteristics and the noise characteristics of a SLA have been measured, and the results have been compared with those predicted by the equivalent circuit model developed in Chapters 5 and 6 as well as with other previous theories proposed by various researchers. These measurements have enabled us to understand more about the principles and operation of SLAs, e.g. the importance of Auger recombination, the effect of beating noise generated by the output of a SLA in a photodiode, the saturation characteristic dependent on the bias current, etc. In addition, it gives some support for the accuracy, applicability and power of the equivalent circuit model proposed by the author in Chapters 5 and 6, which encourages further use of this theoretical model in designing and analysing SLAs with more complicated structures for use in future optical fibre communications systems.

7.7 REFERENCES

[1]   T. Saitoh and T. Mukai, "Recent progress in semiconductor laser amplifiers", *IEEE J. Lightwave Technol.*, **LT-6**, No. 11, pp. 1156–1164, 1988.

[2]   G. P. Agrawal and N. K. Dutta, *Long-wavelength Semiconductor Lasers*, Van-Nostrand Reinhold, New York, 1986.

[3]   H. Kressel, *Semiconductor Devices for Optical Communications*, 2nd edition, Springer-Verlag, Berlin, 1982.

[4]   T. Li, *Topics in Lightwave Transmission System*, Academic Press, London, 1991.

[5]   Y. Imai, E. Sano and K. Asai, "Design and performance of wideband GaAs MMICs for high-speed optical communication systems", *IEEE Trans. Microwave Theory Technol.*, **40**, No. 2, pp. 185–189, 1992.

[6]   A. Yariv, *Optical Electronics*, 3rd edition, Holt–Saunders, 1985.

[7]   J. M. Senior, *Optical Fibre Communications: Principles and Practice*, 2nd edition, Prentice-Hall, New York, 1992.

[8]   T. Saitoh and M. Mukai, "1.5 µm GaInAs travelling-wave semiconductor laser amplifier," *IEEE J. Quantum Electron.*, **QE-23**, No. 6, pp. 1010–1020, 1987.

[9]   Y. Yamamoto, *Coherence Amplification and Quantum Effects in Semiconductor Lasers*, John Wiley and Sons, New York, 1991.

[10] R. Olshansky, C. B. Su, J. Manning and W. Powazink, "Measurement of radiative and nonradiative recombination rates in InGaAsP and AlGaAs light sources", *IEEE J. Quantum Electron.*, **QE-20**, No. 8, pp. 838–854, 1984.

[11] M. J. Adams, "Time dependent analysis of active and passive optical bistability in semiconductors", *IEE Proc.*, **132**, Part J, No. 6, pp. 343–348, 1985.

[12] C. Y. J. Chu and H. Ghafouri-Shiraz, "Structural effects on polarization sensitivity of travelling-wave semiconductor laser amplifiers", *3rd Bangor Communication Symposium*, United Kingdom, pp. 19–22, 1991.

[13] H. Ghafouri-Shiraz, "A model for peak-gain coefficient in InGaAsP/InP semiconductor laser diodes", *Opt. Quantum Electron.*, **20**, pp. 153–163, 1988.

[14] J. Wang, H. Olesen and K. E. Stubkjaer, "Recombination, gain and bandwidth characteristics of 1.3 μm semiconductor laser amplifiers", *IEEE J. Lightwave Technol.*, **LT-5**, No. 1, pp. 184–189, 1987.

[15] I. D. Henning, M. J. Adams and J. V. Collins, "Performance prediction from a new optical amplifier model", *IEEE J. Quantum Electron.*, **QE-21**, No. 6, pp. 609–613, 1985.

[16] H. Ghafouri-Shiraz, Unpublished work, 1992.

[17] A. Yariv and P. Yeh, *Optical Waves in Crystals*, John Wiley and Sons, New York, 1984.

[18] R. M. Jopson, G. Eisenstein, K. L. Hall, G. Raybon, C. A. Burrus and U. Koren, "Polarisation-dependent gain spectrum of a 1.5 μm travelling-wave optical amplifier", *Electron. Lett.*, **22**, No. 21, pp. 1105–1107, 1986.

[19] A. W. Snyder and J. D. Love, *Optical Waveguide Theory*, Chapman & Hall, London, 1983.

[20] J. John, T. S. M. Maclean, H. Ghafouri-Shiraz and J. Niblett, "Matching of single-mode fibre to laser diode by microlenses at 1.5 μm wavelength", *IEE Proc.*, **141**, Part J, No. 3, pp. 178–184, 1994.

[21] H. Ghafouri-Shiraz and C. Y. J. Chu, "Effect of phase shift position on spectral linewidth of the π/2 distributed feedback laser diode", *J. Lightwave Technol.*, **LT-8**, No. 7, pp. 1033–1038, 1990.

[22] D. Marcuse, "Computer model of an injection laser amplifier", *IEEE J. Quantum Electron.*, **QE-19**, No. 1, pp. 63–73, 1983.

[23] H. Ghafouri-Shiraz, "Single transverse mode condition in long wavelength SCH semiconductor laser diode", *Trans. IEICE Jpn*, **E70**, No. 2, pp. 130–134, 1987.

[24] C. Vasallo, "Rigorous and approximate calculations of antireflection layer parameters for travelling-wave diode laser amplifiers", *Electron. Lett.*, **21**, No. 8, pp. 333–334, 1985.

[25] C. Vassallo, "Theory and practical calculation of antireflection coatings on semiconductor laser diode optical amplifiers", *IEE Proc.*, **137**, Part J, No. 4, pp. 193–202, 1990.

[26] Y. Mukai, Y. Yamamoto and T. Kimura, "Optical amplification by semiconductor lasers", *Semiconductors and Semimetals*, **22**, Part E, pp. 265–319, Academic Press, London, 1985.

[27] T. Mukai and T. Yamamoto, "Noise characteristics of semiconductor laser amplifiers", *Electron. Lett.*, **17**, No. 1, pp. 31–33, 1981.

[28] T. Mukai and Y. Yamamoto, "Noise in an AlGaAs semiconductor laser amplifier", *IEEE J. Quantum Electron.*, **QE-18**, No. 4, pp. 564–575, 1982.

[29] J. A. Armstrong and A. W. Smith, "Intensity fluctuation in GaAs laser emission", *Phys. Rev.*, **140**, No. 1A, pp. A155–A164, 1965.

[30] H. Taub and D. L. Schilling, *Principles of Communication Systems*, 2nd edition, McGraw-Hill, New York, 1986.

[31] C. Y. J. Chu and H. Ghafouri-Shiraz, "A simple method to determine carrier recombinations in a semiconductor laser optical amplifier", *IEEE Photon. Technol. Lett.*, **5**, No. 10, pp. 1182–1185, 1993.

[32] C. Y. J. Chu and H. Ghafouri-Shiraz, "Analysis of gain and saturation characteristics of a semiconductor laser optical amplifier using transfer matrices", *IEEE J. Lightwave Technol.*, **LT-12**, No. 8, pp. 1378–1386, 1994.

8

CONCLUSIONS

8.1 SUMMARY OF THE BOOK

In this book, the performance characteristics of semiconductor laser amplifiers (SLAs) have been studied both theoretically and experimentally. As discussed in the introductory Chapter 1, these laser amplifiers can be used in optical fibre communication systems as in-line repeaters or they can be integrated with the optical receiver chip as an optical pre-amplifier. The basic principles of optical amplification have been explored in Chapter 2 by considering the interaction of electromagnetic radiation with a two-level quantum mechanical system. This has been shown to give a qualitatively accurate picture of the principles and performance characteristics of all types of optical amplifier, including both SLAs and fibre amplifiers. The discussion has been narrowed down to SLAs in Chapter 3 by considering in greater detail how semiconductor lasers have been used as optical amplifiers, both by examining their physical principle of operation as well as reviewing the historical development of using semiconductor lasers as optical amplifiers. In this discussion, SLAs have been classified by their structural differences (i.e. Fabry–Perot amplifiers (FPAs) and travelling-wave amplifiers (TWAs)). Also the subsequent differences in their performance characteristics have been examined. Typical applications of SLAs as in-line repeaters, pre-amplifiers, and simple routing switches have also been explored.

A systematic theoretical study of the performance characteristics of SLAs is the major objective in Chapters 4–6. The SLAs have been treated as an active dielectric optical waveguide in Chapter 4, in which the propagation characteristics of optical signals have been analysed using electromagnetic theory. The resulting modal gain coefficient of the SLA, which determines the single-pass gain and hence the overall gain of the amplifier for a particular input signal wavelength, has been analysed using a perturbation technique proposed by Ghafouri-Shiraz and Chu [1]. This technique allows the polarisation sensitivity of SLAs with a simple buried heterostructure (BH), the most common SLA structure, as well as more complex structures like stripe loaded structures and buried channel (BC) structures, to be analysed quickly and accurately. Furthermore, the modal gain coefficients for SLAs with more complicated transverse modal field distributions (e.g. rib structures) can also be calculated. For these structures, a single optical confinement factor, as used commonly in index-guiding BH SLAs, cannot be used to calculate the modal gain coefficients directly. Often, several similar "optical confinement factors" have to be defined in each region in the transverse plane before the modal gain coefficients can be calculated [1]. Hence the proposed technique provides an alternative and yet effective way to calculate the modal gain coefficient directly without involving conventional formulae with optical confinement factors. The structural design for polarisation insensitive SLAs has been explored, showing that a thick active layer will minimise the polarisation sensitivity of amplifiers with a BH. It can be minimised further by using alternative structures such as BC structures. Incidentally, BC structures become more and more common for the integration of electronic and optical devices. The importance of maintaining single-transverse mode (STM) operation in SLAs has also been discussed in this chapter.

The gain characteristics and the corresponding saturation behaviour of SLAs have been analysed in Chapter 5. The analysis is a complex one as there are continuous interactions between the photons and injected carriers due to the bias current, both in a temporal and a spatial sense. The temporal relation between the photons in the SLA cavity and the injected carrier density is governed by a set of rate equations, which form the basis of the analysis of SLAs. The spatial relation between the photons and the injected carrier density can be included in the analysis by introducing travelling-wave equations to analyse the variation of signal field and the intensity of spontaneous emissions inside the SLA. The solution procedures can be very tedious even for simple structures such as BH FPAs and TWAs. An approximation based on averaging the photon density along the SLA has been examined, which will simplify the solution procedures. However, both the exact and the approximate techniques require the field distribution in the SLA cavity to be solved analytically before numerical iterations are applied to solve the equations. For more complex structures (e.g. two-section amplifiers, SLAs with a tapered active region, etc.), this may not be possible. A robust and efficient technique based on transfer matrix analysis, as proposed by Chu and Ghafouri-Shiraz [2, 3], has been described in Chapter 5. This technique does not require any averaging approximation and will be able to account for the spatial relationship between photon density and carrier density along the amplifier, and has the advantage of ease of implementation on computers and greater flexibility compared with the approximation technique. The application of the proposed technique for analysing the effects of stray reflections from fibre ends on the gain characteristics have been demonstrated in a later chapter. The effects of the structural parameters on the gain characteristics of BH SLAs have also been studied, and the implications for system design have been discussed briefly.

The effects of spontaneous emissions have been neglected in the discussion of Chapter 5. In Chapter 6 spontaneous emissions are first examined by using the photon statistics master equations approach, the rate equation approach and the travelling-wave equation approach. From these basic investigations, it has been found that the spontaneous emissions from SLAs generate additional beat noises when their output is detected by photo-diodes. Then the technique explained in Chapter 5 has been extended to construct a full equivalent circuit model for analysing SLAs. It has been seen that any SLA can be modelled by a cascaded network of negative active resistors and lossless transmission lines. By terminating the network with an appropriate load impedance, and by changing the characteristic impedance of the transmission lines, any refractive index steps occurring along the amplifier can be modelled. This allows complex structures to be analysed. The resulting noise due to quantum mechanical fluctuations in the negative resistors is found to be both qualitatively and quantitatively accurate for modelling the effects of spontaneous emissions in actual SLAs. This has been justified by comparing the deduced spontaneous emission power emitted in a SLA without input signal, using the equivalent circuit model, with that calculated by Green's function. Therefore simple circuit analysis techniques supplemented by some knowledge of signal analysis can be used to compute the noise generated in SLAs, rather than involving the more complicated concepts of Langevin forces and statistical physics. It can also take into account the spatial dependence of the photon density and carrier density along the SLA cavity. The power of the equivalent circuit model has been illustrated by analysing the effects of stray reflections from fibre ends on the noise characteristics of SLAs. Finally, the structural design of low-noise BH SLAs has been discussed, and the effects of noise on the design of optical communication systems with SLAs as in-line repeaters and pre-amplifiers has been examined.

Any theoretical study on SLAs needs to be supported by a complementary experimental study. This has been described in Chapter 7, where the experimental set-up, and the principles of SLA measurements and experimental procedures have been described in detail. Three particular aspects of SLAs have been studied experimentally. First, the recombination mechanisms of injected carriers in SLAs have been investigated by measuring the amplified spontaneous emission power of the SLA, and it has been found that non-radiative recombinations, such as Auger recombinations, in the $1.5\,\mu m$ wavelength region are extremely important. In fact, these recombination mechanisms allow the structure of the active region to be tailored to achieve STM operation, with a relatively thick active layer to reduce polarisation sensitivity. In SLAs with materials for the optical amplification of shorter wavelengths, where non-radiative recombinations are not important, such dimensions of the active region will result in a lower threshold current for oscillation, hence limiting the range of injection current which is allowed to flow through the SLA, as illustrated in Chapter 5. Secondly, the gain characteristics were measured for both TE and TM polarisations of the input optical signal. The low polarisation sensitivity claimed by the supplier has been observed. The effects of the output power level and the injection current on the overall gain and output saturation power of the SLA have been examined. These results have been compared with the theoretical predictions made by the equivalent circuit model [2, 3] and the approximation method, and these illustrate the importance of the spatial dependence of photon density and carrier density in the SLA. Finally, the noise current generated in an InGaAsP pin photodiode by a SLA with different input optical power levels has been measured.

8.2 LIMITATIONS OF THE RESEARCH STUDY

The strength of any scientific or engineering model cannot be fully utilised without being aware of its limitations. Similarly, the significance of any research cannot be fully appreciated without examining its limitations. In this section we examine some of the limitations in this SLA study. Three particular aspects will be examined: the limitations of the analytical models used in this book, the limitations of the experimental studies and the limitations in the scope of the overall research work presented here.

8.2.1 Limitations on theoretical studies

There are mainly two techniques used in this book. In Chapter 4, the analysis of modal gain coefficients in SLAs by perturbation techniques has been examined. In Chapters 5 and 6, an equivalent circuit model has been developed for analysing SLAs and this can be implemented on computers using transfer matrix methods. To analyse a SLA, whatever its structure, both of these techniques should be combined to form a systematic framework for this purpose.

The discussion in Chapters 4–6 may lead the reader to conclude that the proposed model may only work for index guided buried heterostructures (BH), because, as we have seen in Chapters 5 and 6, the optical confinement factor has been used to calculate the modal gain coefficient in the SLA, which is an essential parameter in solving for the coefficients of the transfer matrices. Indeed, the analysis and discussion have been limited to BH SLAs because they are the most common structures for SLAs at present. However, there is no reason why we cannot extend the framework to analyse other structures for which a single optical confinement factor cannot be used to calculate the modal gain coefficient (e.g. buried channel structures, rib structures, DFB structures, etc.). The only modification required is to replace the modal gain coefficients in the TMM analysis with the perturbation expression discussed in Chapter 4, which does not require the use of an optical confinement factor in the calculation.

The real limitation of the model is that, because of the way the equivalent circuit has been constructed, it will only work for index guided structures. For gain guided structures, the longitudinal propagation constant β_z is not independent of z, nor the transverse positions x and y. In principle, however, it may be possible to modify the framework of analysis following the arguments of Buus [4]. Nevertheless, most of the SLAs and laser diodes manufactured now are index guided. This is because the coupling efficiency to optical fibres is better for the index guided laser structure [5]. There are recent trends, especially for DFB lasers, in using gain guided structures to stabilise the single longitudinal mode (SLM) operation under direct modulation [6]. However, for the purpose and scope of the present book, concentrating on index guided structures will be sufficient.

8.2.2 Limitations on experimental studies

To be in line with theoretical studies, the experimental measurements discussed in Chapter 7 concentrate only on the gain, gain saturation and noise characteristics of SLAs. The complexity involved in the experimental arrangement required to perform tests in simple optical communication systems, or to measure characteristics such as cross-talk, harmonic distortion, gain saturation in amplification of short optical pulses, etc., will be significantly higher. The major problem encountered in the measurements, as discussed in Chapter 7, is the need to have very good alignment between optical components in order to maintain a stable and optimum coupling. Compared with fibre amplifiers, SLAs indeed suffer from this disadvantage of relative difficulty in coupling. This problem was greater in this case because the SLA used in the experiment was an unpacked type. It will be necessary, in more complex experiments, to have SLAs that are packaged if reliable and accurate measurements are to be made.

8.2.3 Limitation on the scope of this book

Finally, a comment about the results presented in this book. The most important performance characteristics, namely gain, gain saturation and noise, have been analysed. The frequency bandwidth has been discussed briefly as the gain spectra of FPAs and TWAs have been compared. This has not been discussed in greater depth because modern SLAs are all coated, and the bandwidth will be significantly closer to that of the material gain spectrum [7]. Furthermore, as we did not consider wavelength division multiplexing (WDM) in this book, the frequency bandwidth in SLAs has not been an important issue. For the same reason, cross-talk, inter-modulation harmonic distortion [8] and four-wave mixing have not been considered. Nor have the amplification of short optical pulses [9, 10] and the switching performance of SLAs been considered. This is not because these are unimportant, but simply that it has not been possible to cover all these topics within the scope of the present book.

8.3 FUTURE WORK

As discussed above, the purpose of this book is to present a new method of analysing the performance characteristics of SLAs. The power of this method has been illustrated by analysing the basic characteristics of simple index guided BH SLAs, and this is supported by experimental measurements. Two new research directions are possible based on the work presented in this book. The first new direction is to extend it to analyse index guided SLAs with more complex structures. This will include rib structures [11, 12], DFB laser amplifiers [13], and two-section or multi-section laser amplifiers [14]. There is also the potential to apply the model to quantum well structures [15–18]. The major differences between quantum well laser amplifiers and bulk laser amplifiers which we have examined so far, will be on the recombination mechanisms [19], dependence of material gain coefficients, refractive index and optical confinement factor on the band structure and injected carriers to the quantum wells [20, 21]. We could replace the models used for bulk structures by those appropriate for quantum wells, and the entire analysis for quantum well laser amplifiers will follow an identical framework to that used for the bulk structures described in this book. In addition, since transfer matrix methods have been readily applied to analyse semiconductor lasers [22–24], the equivalent circuit model may also be extended to model semiconductor lasers as well. Together with other models used to analyse passive optical devices [25] and active electronic components [26], this may form the basis of a general computer-aided design package to design integrated opto-electronic devices [27].

The second new direction is to use the equivalent circuit model to investigate other performance characteristics of SLAs. Three specific areas can be pursued further using the present model without significant modifications. The first area will be to extend the analysis of noise to consider the effect of spontaneous emissions of SLAs on the linewidth of the signal [28–30] and the resulting phase noise introduced into the system [31]. This can be used to consider the bit error rate of the system in more detail than has been considered in this book [32, 33]. The second area will be in the analysis of analogue intensity modulated systems in the radio frequency (RF) and microwave frequency regions. In such applications the harmonic distortion introduced by the non-linearity in the SLAs becomes important. Finally, the amplification of short optical pulses can be examined by this method. Such analysis will be useful for predicting the performance limits of SLAs used for the high-speed digital transmission in optical fibre communication systems.

8.4 REFERENCES

[1] H. Ghafouri-Shiraz and C. Y. J. Chu, "Analysis of waveguide properties of travelling-wave semiconductor laser amplifiers using perturbation techniques", *Fiber and Integrated Optics*, **11**, pp. 51–70, 1992.

[2] C. Y. J Chu and H. Ghafouri-Shiraz, "Equivalent circuit theory of spontaneous emission power in semiconductor laser optical amplifiers", *IEEE J. Lightwave Technol.*, **LT-12**, No. 5, pp. 760–767, May 1994.

[3] C. Y. J. Chu and H. Ghafouri-Shiraz, "Analysis of gain and saturation characteristics of a semiconductor laser optical amplifier using transfer matrices", *IEEE J. Lightwave Technol.*, **LT-12**, No. 8, pp. 1378–1386, August 1994.

[4] J. Buus, "The effective index method and its application to semiconductor lasers", *IEEE J. Quantum Electron.*, **QE-18**, No. 7, pp. 1083–1089, 1982.

[5] J. John, T. S. M. Maclean, H. Ghafouri-Shiraz and J. Niblett, "Matching of single-mode fibre to laser diode by microlenses at $1.5\,\mu m$ wavelength", *IEE Proc.*, **141**, Part J, No. 3, pp. 178–184, 1994.

[6] Y. Yamamoto, *Coherence Amplification and Quantum Effects in Semiconductor Lasers*, John Wiley and Sons, New York, 1991.

[7] G. Eisenstein, "Semiconductor optical amplifiers", *IEEE Circuits and Devices Mag.*, pp. 25–30, 1989.

[8] C. D. Zaglanakis and A. J. Seeds, "Computer model for semiconductor laser amplifiers with RF intensity-modulated inputs", *IEE Proc.*, **139**, Part J, No. 4, pp. 254–262, 1992.

[9] H. F. Liu, M. Tohyama, T. Kamiya and M. Kawahara, "Gain saturation of a 1.3 μm InGaAs travelling-wave amplifier in picosecond pulse amplification", *Jpn J. Appl. Phys.*, **31**, Part 2, No. 4B, pp. L487–L489, 1992.

[10] A. Uskov, J. Mørk and J. Mark, "Theory of short-pulse gain saturation in semiconductor laser amplifiers", *IEEE Photon. Technol. Lett.*, **4**, No. 5, pp. 443–446, 1992.

[11] I. Maio, "Gain saturation in travelling-wave ridge waveguide semiconductor laser amplifiers", *IEEE Photon. Technol. Lett.*, **3**, No. 7, pp. 629–631, 1991.

[12] N. Storkfelt, B. Mikkelsen, D. S. Olesen, M. Yamaguchi and K. E. Stubkjaer, "Measurement of carrier lifetime and line width enhance factor for 1.5 μm ridge-waveguide laser amplifier", *IEEE Photon. Technol. Lett.*, **3**, No. 7, pp. 632–634, 1991.

[13] T. Makino and T. Glinski, "Transfer matrix analysis of the amplified spontaneous emission of DFB semiconductor laser amplifiers", *IEEE J. Quantum Electron.*, **QE-24**, No. 8, pp. 1507–1518, 1988.

[14] R. M. Fortenberry, A. J. Lowery and R. S. Tucker, "Up to 16 dB improvement in detected voltage using two section semiconductor optical amplifier detector", *Electron. Lett.*, **28**, No. 5, pp. 474–476, 1992.

[15] P. J. Stevens and T. Mukai, "Predicted performance of quantum-well GaAs-(GaAl)As optical amplifiers", *IEEE J. Quantum Electron.*, **QE-26**, No. 11, pp. 1910–1917, 1990.

[16] K. S. Jepsen, B. Mikkelsen, J. H. Povlsen, M. Yamaguchi and K. E. Stubkjaer, "Wavelength dependence of noise figure in InGaAs/InGaAsP multiple-quantum-well laser amplifier", *IEEE Photon. Technol. Lett.*, **4**, No. 6, pp. 550–553, 1992.

[17] K. Komori, S. Arai and Y. Suematsu, "Noise study of low-dimension quantum-well semiconductor laser amplifiers", *IEEE J. Quantum Electron.*, **QE-28**, No. 9, pp. 1894–1990, 1992.

[18] D. Tauber, R. Nagar, A. Livne, G. Eisenstein, U. Koren and G. Raybon, "A low-noise-figure 1.5 μm multiple-quantum-well optical amplifier", *IEEE Photon. Technol. Lett.*, **4**, No. 3, pp. 238–240, 1992

[19] G. P. Agrawal and N. K. Dutta, *Long-wavelength Semiconductor Lasers*, Van-Nostrand Reinhold, New York, 1986.

[20] A. Yariv, *Quantum Electronics*, 3rd edition, John Wiley and Sons, New York, 1989.

[21] C. Aversa and K. Iizuka, "Gain of TE–TM modes in quantum-well lasers", *IEEE J. Quantum Electron.*, **QE-28**, No. 9, pp. 1864–1873, 1992.

[22] G. Bjork and O. Nilsson, "A new exact and efficient numerical matrix theory of complicated laser structures: properties of asymmetric phase-shifted DFB lasers", *IEEE J. Lightwave Technol.*, **LT-5**, No. 1, pp. 140–146, 1987.

[23] G. Bjork and O. Nilsson, "A tool to calculate the linewidth of complicated semiconductor lasers", *IEEE J. Quantum Electron.*, **QE-23**, No. 8, pp. 1303–1313, 1987.

[24] M. G. Davis and R. F. O'Dowd, "A new large-signal dynamic model for multielectrode DFB lasers based on the transfer matrix method", *IEEE Photon. Technol. Lett.*, **4**, No. 8, pp. 838–840, 1992.

[25] C. H. Henry and Y. Shani, "Analysis of mode propagation in optical waveguide devices by Fourier expansion", *IEEE J. Quantum Electron.*, **QE-27**, No. 3, pp. 523–530, 1991.

[26] Y. Imai, E. Sano and K. Asai, "Design and performance of wideband GaAs MMICs for high-speed optical communication systems", *IEEE Trans. Microwave Theory Technol.*, **40**, No. 2, pp. 185–189, 1992.

[27] U. Koren, R. M. Jopson, B. J. Miller, M. Chien, M. C. Young, C. A. Burrus, C. R. Giles, H. M. Presby, G. Raybon, J. D. Evankow, B. Tell and K. Brown-Goebeler, "High power laser-amplifier photonic integrated circuit for a 1.48 μm wavelength operation", *Appl. Phys. Lett.*, **59**, No. 19, pp. 2351–2353, 1991.

[28] M. J. Connelly and R. F. O'Dowd, "Designing optically repeated links using a new semiconductor laser amplifier noise model", *Opt. Amplif. Applic.*, Topical Meeting of OSA, **13**, pp. 184–187, 1990.

[29] M. J. Connelly and R. F. O'Dowd, "Theory of signal degradation in semiconductor laser amplifiers with finite facet reflectivities", *IEEE J. Quantum Electron.*, **QE-27**, No. 11, pp. 2397–2403, 1991.

[30] K. Hinton, "Optical carrier linewidth broadening in a travelling wave semiconductor laser amplifier", *IEEE J. Quantum Electron.*, **QE-26**, No. 7 pp. 1176–1182, 1990.

[31] K. Kikuchi, C. E. Zah and T. P. Lee, "Measurement and analysis of phase noise generated from semiconductor optical amplifiers", *IEEE J. Quantum Electron.*, **QE-27**, No. 3, pp. 416–422, 1991.

[32] P. A. Humblet and M. Azizoglu, "On the bit error rate of lightwave systems with optical amplifiers", *IEEE J. Lightwave Technol.*, **LT-9**, No. 11, pp. 1576–1582, 1991.

[33] D. Marcuse, "Calculation of bit-error probability for a lightwave system with optical amplifiers", *IEEE J. Lightwave Technol.*, **LT-9**, No. 4, pp. 505–513, 1991.

INDEX

underlined italic references are to tables
underlined references are to figures

absorption 12–14, 36
active layer thickness
  gain, BH 90–1, *92*, 137
  noise figure 179, *180*, *181*
  optical confinement 87, *89*
active layer width
  gain 135, *136*
  modal transverse field distribution 96–8
  optimisation 98
active region 34, *35*
  dimensions, experimental *192*, 211
  modes 36
alignment 216, 227
amplification
  parametric, noise figure 21
  by pumping 17
  in stimulated emission 12
  and thermal equilibrium 13
amplified spontaneous emission: *see* ASE
amplifier 6
  bandwidth 18–19
  characterisation 16–22
  cleaved substrate 45–6
  FPA limitations 107
  heterostructure 46–8
  homojunction 44–6
  ideal 22–3
  laser as 34
  noise figures 20–2
  performance characteristics 9–10
  performance requirements 22–3, *56*
  phase-insensitive 25–6, 28
  practical 23–9
  qualitative criteria 23–9
    bandwidth 24
    noise figure 25–9
    signal gain 24

  quantum well 228
  saturation output power 19–20
  signal gain 16–18
  SLD 44–8
  see also pre-amplifier; SLA
amplifier cascade, noise figure 21–2, 183
amplifier cascade power 139
amplifier gain: *see* gain
amplifier saturation, ASE 58
anti-reflection (AR) coating 49–50
  and bandwidth 227
  gain 139
  homojunction 45
  reflectivity window 51
ASE (amplified spontaneous emission) 20–2
  amplifier chaining 22
  backwardly emitted 22
  heterostructure 47
  phase contamination 22
  repeater chain saturation 58
aspect ratio, waveguide 97
attenuation, fibres 1, *2*, 182–3
attenuation factor 15
Auger recombination 201
  experimental results 203–4
  TMM 128
AVPD (averaged photon density) 143
  experimental studies 203–4
  FPA analysis 128–9
  gain calculation 211, *213*
  TMM compared 129–30
    experimental 214

band tailing 41, 53
bandwidth 1–2, *3*, 18–19
  amplifier detector 23
  coating 227

bandwidth *(cont.)*
  data rate 1
  laser coherency 2
  performance limits 24
  pre-amplifier *56*, 58
  resonance 19
  ripples 49
  for switching 60–1
  TWA/FPA 50
  see also material gain bandwidth
beam-splitter 204
beat noise 153, *176*
  noise measurement 216
  shot noise comparison *176*, 177
  signal/spontaneous emission 154, 155, *218*, 219
  spontaneous emission 153, 154, 155, 156
  suppression *175*, 177
beat noise power 216
BH (buried heterostructure) SLA
  active layer thickness gain 90–1, *92*, 137
  material gain 86
  modelling 121, 121–5
  refractive index, TWA *91*
  single pass gain layer thickness 90–1, *92*
  STM 87
bias
  laser 192, *193*
  output power 204, 205
  SLA 199
  spontaneous emission 126
bias current: *see* injection current
biasing (forward), and population inversion 34, *35*
boundary, rectangular waveguide 75

cascade, power losses 139
carrier, effect on noise figure 179–81
carrier density
  and bias current 38–9
  fluctuation 149–50
  FPA distribution *131*
  and gain 34–5
  and material gain coefficient 37–8
  matrix methods 117
  and optical field 126
  photon density dependence 114
  photon number relation 157
  rate equation analysis 114
  spread 37
carrier recombination 104

carrier recombination rate 39
cavity length, and spectral behaviour 143
cavity noise enhancement, FPA 156
cavity resonance
  AVPD 130
  bistable switching 59–60
  facet reflectivity 49
  FPA 154
  gain 135, 139
  length 135
  TWA/FPA 50
  see also resonance
cladding, leakage into 36
cladding loss, and perturbation analysis 86–7
clamping, peak gain 43
cleaved substrate amplifier *52*, 53
coating: *see* anti-reflection coating
coherence
  population inversion 36
  stimulated emission 12
coherent detection 3–4, *5*
communication, optical xi–xii
communication system
  configuration 2, *4*
  experimental studies difficulties 227
  IM, system building 139–41
  speed xi
  system considerations 182–3
confinement, and directional waveguides 97
continuous wave (cw) operation, homojunction 44–5
conversion, optical/electronic xii
core, rectangular waveguide 75
core power confinement 81
coupling, signal 206, *207*
coupling efficiency 139
coupling losses: *see under* loss
curl equations 68–9
current pumping: *see* pumping
current source, for laser/SLA 192, *193*

dark current, noise measurement 216–17
decay rate 12–13
decaying parameters 76, 79, 80
degeneracy 13
depletion electric field 34
detection systems 2–5
detector, amplifier as 23
detuning, SLA switch 59–60
dielectric constants, perturbed/unperturbed 85–6

diffusion, degenerate semiconductors 34
diffusion coefficient, TWA 184–5
dispersion, in amplifier 18, 19; *see also* wave-
    guide
dispersion curves, rectangular waveguide 80–2
dopant, waveband selection 2

effective gain: *see* modal gain coefficient
effective index method: *see* EIM
efficiency, photon statistics gain formulation
    109
eigen equations 73–4
    guided mode equation solution 72–3
    quasi-TM mode 79
EIM (effective index method) 78–82
    disadvantages 80
    Marcatilli's solutions compared 80
Einstein's relations 13
electric field
    depletion 34
    matrix analysis *119*
    output (amplifier) 25
    quadrature representation 25–6, *27*
    TWA 184
electric field distribution, FPA 112
electric field wave equation 69, 70
energy bands 53; *see also* band tailing
energy density, spectral 13
energy levels, population distribution 16
energy transition: *see* transition
ensemble averaging 26, 158
envelope detection 2–3
equivalent circuit modelling
    applicability 226–7
    index guided structure 226
    noise 159–75
    power 167
    SLA 130, 132–5, 144
    spontaneous emission 163–70
    sub-networks 163–4
erbium 2, 6, 9
evanescent waves 71
excess noise coefficient 156
excited state, population density 11
extraction efficiency 109–10

FA (fibre amplifiers) 6, 7, 9
Fabry–Perot amplifier: *see* FPA
Fabry–Perot cavity model *105*
facet, feedback from 48
facet loss, modelling 171

facet modelling 134
facet reflectivity: *see* reflectivity
facet reflectivity measure 107
feedback 17
    end-facets 48
    oscillation 34
Fermi energy levels 34
fibre amplifiers: *see* FA
field distributions 67
field profile, longitudinal, FPA *131*
filter
    optical 61
    tunable 5
flip-flop: *see* switch, bistable
FM demodulation 61
FPA (Fabry–Perot amplifier) 17, 48
    analysis model *105*
    frequency mismatch sensititivity 107–9
    gain 167
    gain determination 105–9
        photon statistics comparison 110
    Green function spontaneous emission power
        184–5
    minimum noise 156
    noise power 185–6
    photon statistics noise analysis 153–6
    spontaneous emission equivalent circuit 163–
        70
    TMM analysis 127–8
    TWA compared 50, 178
frequency
    carrier density 43
    material gain coefficient 41, *42*
    see also wavelength
frequency bandwidth: *see* bandwidth
frequency modelling 120
    spontaneous emissions 161
    see also noise
frequency spread, transition 15

gain 33–4
    analysis assumptions 104
    AVPD 211
    carrier density 34–5
    change by structural effects 135–9
    components 168
    dependence on propagation constant 106
    experimental measurement 210–11
    FPA with uniform profile 167
    frequency 18; *see also* bandwidth
    impedance modelling 160

gain *(cont.)*
  infinite 17–18
  maximum/minimum 106
  oscillation 24
  output power *129*
  performance limits 24
  photon statistics formulation 109–10, 152
  polarisation 36, 210–11
  reflections effect 37, 137, *138*, 139, *174*, 175
  and stray reflections *173*
  see also material gain; single pass gain
gain analysis, amplifiers 37–44
gain characteristic measurement 204–14
gain coefficient: *see* material gain coefficient
gain guided structure
  equivalent circuits 226–7
  modal field solution 87
gain profile
  non-uniform, analysis by TMM 125–7
  TW equations 116–17
  uniform 112–13
gain saturation: *see* saturation
gain spectra *42*, 43
  heterostructures 47
  and noise spectrum *151*
gain spectrum broadening 43
gain spectrum ripples, FPA 106–7, *108*; *see also*
  ripples
gain uniformity
  heterostructures 46
  homojunctions 46
gain value, ideal amplifier 23
Green function 162, 170
  spontaneous emission power 183–6
guided mode 71, *72*
  field solution 72–3
  photon density 15

heat pump, for temperature control 195–7, 199
heterodyne analogy, coherent detection 3–4
heterostructure
  buried 52
    modal gain coefficients 84–7
  separate confinement 52–3
  see also BH
heterostructure amplifiers 46–8
homojunction 34
  amplifiers 44–6
hybrid modes 76
  active layer width 96–8
  refractive index distribution 94–6

  see also TE (quasi); TM (quasi)
hysteresis, bistable switching 60

IM (intensity modulation) system 4–5, 139–41
  RF/microwave 228
impedance, for gain modelling 160
in-line repeater: *see* repeater
index guided structure
  complex structure 227–8
  equivalent circuit model 226
  gain components 168
  perturbation analysis 84–7
induced absorption 35
injected carrier: *see* carrier
injection (bias) current
  gain *212*, *213*
  injected carrier density 38–9
  modal gain coefficient 201
  noise *219*, 220
  peak carrier density 39
injection efficiency 109
input, noise contamination 27, 28
input power, true 210
integrated circuits xii
intensity modulation: *see* IM
isolator 204, *205*, *207*

laser, as amplifier 34
laser, experimental
  biasing 192, *193*
  current source 192–3
  pulsed operation 193–4
  temperature stabilisation 194–7, 199
lasing 142
  and oscillation 24
  photon number gain 141
  stimulated emission 13–14
length of amplifier
  gain 135, *136*
  noise figure 179, *180*, *181*
light pumping 17; *see also* FA
light–electrical conversion xii, 58
linewidth studies, SLA 228
loads, as facet models 134
longitudinal propagation constant 70–1
loss 5
  communication systems 139–40
  component number 216
  coupling 227
    experimental determination 209–10
    photon statistics gain formulation 109
  modelling by mirror 171

see also signal-to-noise ratio
loss coefficient   36

magnetic field vector wave equation   69, 70
Marcatilli's waveguide analysis   75–8
Marcatilli/EIM solutions compared   80
material gain   111, 111–12
  and transition probability   55
material gain bandwidth   18–19
material gain coefficient   20, 36
  approximations   37–44
  frequency   41, *42*
  modal gain coefficient   201
  peak   37–8, 39, 202
    clamped   43
  power   37
  practical determination   104
  rate equation analysis   114
  single pass gain   83
  spectral function expression   41, *42*
  spontaneous emission fluctuations   169
  stimulated emission   37
material gain profile
  longitudinal field distribution   120
  TMM   128
material gain spectra, and bandwidth   24
material gain wavelength, peak   44
matrix, three-dimensional modelling   120
matrix methods   117–20
microwave engineering/SLA analogy   132–5
mirror
  asymmetrical, and noise figure   179, 180–1
  as reflection model   170–1
modal field (transverse), by EIM   78–82
modal field (transverse), solution   71–2
  Bessel function   82
  finite element method   82
  integral equation approach   83
  iterative equivalent index method   82
  rectangular waveguide   75–8
  three-layer slab   68–75
  weighted index method   83
modal field distribution   68
modal gain coefficient   83
  applicability   226
  calculation   84–7
  and injection current   201
mode   36, 68
  continuous radiating   71, *72*
  odd/even   73, 74
  and optical confinement   90

propagation constant   74
side modes   45
TE/TM   70
see also guided mode; hybrid mode
mode discrimination, waveguide parameters   76
modulation methods   2, *5*
modulator
  amplifier as   23
  for pulsed laser   192, *194*
  SLA   61
momentum, statistical   150
monitoring networks   61
multiplexing   1
  gain saturation   20
  regenerative repeaters   57

N-port network   118
noise
  and bias current   *219*, 220
  equivalent circuit modelling   159–75
    spontaneous emissions   160–2
    stray reflections   170–5
    validity   162–70
  formulation in SLA   150–9
    photon statistics   150–7
    rate equation   157–9
    travelling-wave equations   159
  Langevin   157–8, 159
    in Green function modelling   183
  photon number fluctuation   158
  physical origins   156–7
  source   149
  TWA/FPA   50–1
  see also beat noise; shot noise
noise analysis, FPA   153–6
noise bandwidth, FPA   156
noise coefficient, excess   156
noise contamination, input   27
noise figure   20–1, 179, 182
  amplifier   20–2
  amplifier chain   183
  amplifier performance limits   25–9
  system considerations   182–3
noise measurements, experimental SLA   199–200
noise minimisation
  design   *175*, 176–82
  for FPA   156
  minimum achievable   29
noise modelling   120, 161
noise performance, FPA/TWA   178

noise power
  FPA 155, 185–6
  operational conditions 181
  output optical power 218
noise power components 176–7
noise power generation 175, *176*

optical amplification
  BH SLA material gain 86
  semiconductor region 34
optical amplification principles 33–44
  gain analysis 37–44
  process in semiconductors 34–7
optical amplifier: *see* amplifier; pre-amplifier
optical communication xi–xii
optical confinement
  active layer thickness 87, *89*
  Poynting and perturbation analysis comparison 87–90
  quasi-TE/M 91–2
  and refractive index 94–6
  slab/hybrid modes 90
optical confinement factor 36–7
  modal gain coefficient 84
optical/electronic conversion xii, 58
optical fibres 1
optical field, carrier density 126
optical filter *208*
optical gain 14–16
  pre-requisites 16
  in TMM analysis 124–5
  see also signal gain
optical power: *see* input p.; output optical p.; power
optical power meter 199
oscillation 18
  ASE 22
  facet feedback 48
  FPA 107
  infinite gain 24
  spontaneous emission 126
  stray reflections 170, 175
oscillation prediction, experimental studies 203–4
oscillator 33
output electric field, and cavity reflections 105–6
output light intensity 19–20
output optical power
  and bias 204, *205*
  effect of reflections *174*

gain *129, 213*
  noise generation, TWA 178
  and noise power 218
  see also saturation output power

peak-gain coefficient model 202, 204
peak-gain wavelength 202
performance, amplifier: *see under* amplifier
perturbation analysis 84–7
  TWA 87–90
perturbation expression, replacing modal gain 226
perturbation and Poynting vector analysis compared 87–90
phase change, and spontaneous emission 150
phase contamination of signal 22
phase delay 132
phase fluctuation, and ASE 25
phase information, and photon emision 115
phase-insensitive linear amplifier 25–6, 28
phase mismatch, and gain 106
phase noise 228
phonons 39
photo-diode 215
  dark current 216–17
photon
  'birth'/'death' analogy 150–1
  carrier density dependence 114
  spontaneous emission, experiment 203–4
photon density 115, 115–16
  frequency band 12
  guided mode 15
  and pumping rate 142
  see also AVPD
photon emission 11
  phase information 115
  stray reflections *172*
photon number
  carrier density relation 157
  fluctuation, and noise 158
  mean 152–3
photon number gain, by amplifier 141
photon population, radiative transitions 152
photon statistics approach
  appraisal 112
  gain formulation 109–10
  noise formulation 150–7
photonic integrated circuits xii
polarisation
  gain 210–11

mode discrimination   76
refractive index distribution   94–6
sensitivity   90–4; *see also under* TWA
waveguide modal discrimination   74
polarisation modes   36
population density
    excited state   11
    thermal equilibrium   13
population density ratio   13
population distribution, energy levels   16
population inversion   11
    carrier density spread   37
    carrier recombination   104
    coherence   36
    for optical gain   16
    photon statistics formulation   152
    by pumping   16
    saturation   20
    semiconductors   34
    stimulated emission   64, 110–11
population inversion parameter   35
    modelling   168
power
    equivalent circuit   167
    input/output signal   28
    and material gain coefficient   37
    optical   15
    see also output optical p.; input p.
power amplification factor   84
power meter: *see* optical power meter
power spectral density (PSD), single-sided   160,
    161, 162
power spectrum modelling   161
power spectrum of spontaneous emissions, and
    material gain   169
Poynting vector and perturbation analysis com-
    pared   87–90
pre-amplifier   5, *56*, 58–9
pre-biasing   193
propagation constant   74
    core power confinement   81
    gain   106
    longitudinal   105
    transverse wave numbers   76
pulsed operation, laser   193–4
pumping
    amplification by   17
    AVPD/TMM compared   130
    negative temperature   16
    saturation   20
pumping rate

photon density   142
stray reflections   *172, 173*

quantum amplifier: *see* amplifier
quantum efficiency, amplifier   169–70
quantum limit operation   25, 28
quantum well laser amplifiers   228
quarter-wave plate   205, *207*
    usage   210–11

radiation   10–11
    overall decay rate   12–13
    spatial dependence   14
    in two-level system   10–16
    see also absorption; spontaneous emission;
        stimulated emission
radiative recombination rate   56
rate equation   143
    analysis   113–14
    noise formulation   157–9
recombination, non-radiative   39;   *see also*
    Auger recombination
recombination coefficient   39
recombination rate   39–40, 55–6
    TMM   128
recombination studies, SLA   200–4
recombination/stimulated emission equilibrium
    111
reflection
    modelled by mirrors   170–1
    refractive index steps   122
    see also anti-reflection coating; facet
reflections (stray)
    effects on spontaneous emission   170–5
    TMM analysis   173–5
reflectivity   17, 51
    amplitude   122, 123
    asymmetrical, and noise figure minimum
        180–1
    cavity resonance   49
    gain   37, 137, *138*, 139, *174*, 175
    mirror models   171
    modelling   134, 135
    noise figure   179, *180, 181*
reflectivity measure   107
reflectivity window, AR coating   51
refractive index   36
    BH TWA   *91*
    stripe geometry TWA   *92*
    TWA buried channel   *93*, 94

refractive index changes, and spontaneous emission 150
refractive index distribution, TWA 94–6
refractive index steps, and reflections 122
repeater 5, 139–41, 182–3
  and ASE 22
  flexibility problems 5–6
  minimum noise figure 25
  non-regenerative 58
  regenerative 5, 56–8
  saturation operation 140
repeater chain saturation 140
residues 82
resistance, negative 132
resonance
  bandwidth 19
  gain 106
resonance mismatch, and gain 106
resonance tuning 107
resonators, spontaneous emission analysis 162
ripple level
  and facet reflectivity 107
  FPA 106–7, 108
ripples 49
  disappearing 50; see also TWA
  and amplifier length 135
  FPA 106–7, 108

saturation 20, 110–12
  amplifier cascade 22
  analysis assumptions 104
  beat noise suppression 175, 177
  heterostructures 47
  homogeneity 112
  homogeneous 43
  mechanisms of 24
  pumping 20
  rate equation analysis 114
  repeater chain 140
  spontaneous emission 112, 173
  stray reflections 170
  TM/TE modes 214
  TWA/FPA 50
saturation analysis, uniform profile 141–3
saturation intensity 111
saturation operation of repeaters 140
saturation output power 19–20, 111, 127, 214
  ideal amplifier 23
  performance limits 24
  see also output optical power
scattering matrix 119

semiconductor, degenerate 34
semiconductor laser amplifiers: see SLA
semiconductors, optical processes 34–7
shot noise 153, 176, 218, 219
  beat noise compared 176, 177
  experimental 216
  spontaneous emission 149, 156
    analogy 160–2
  thermal 160
side modes, homojunctions 45
signal attenuation: see attenuation
signal coupling 206, 207
signal fields, matrix methods 117
signal gain 16–18; see also optical gain
signal-to-noise ratio 21
  amplifier 28
  communications systems 182, 183
  in envelope detection 3
  IM systems 4–5
  improving 5
single pass gain
  active layer width 97
  defined 17
  layer thickness, BH 90–1, 92
  material coefficient 83
  optical confinement 95, 96
  TWA buried stripe 92
single transverse mode: see STM
SLA (semiconductor laser amplifiers) 6–7, 9
  active layer thickness, and optical confinement 87, 89
  applications 61; see also specific application
  electromagnetic analysis 67–8
  equivalent circuit model 130, 132–5
  noise minimisation criteria 175, 176–82
  operational classification 48–51
  performance limit prediction 228
  schematic 49
  structural classification 51–3
  transmission experiments 47–8
  uniform gain profile 104–13
SLA, experimental
  bias 199
  detection circuit 199–200
  gain characteristics measurement 204–14, 210–11
    coupling losses 209–10
    experimental set-up 204–9
    results 211–14
  noise characteristics 214–20
    experimental set-up 214–15

method 215–17
  results 217–20
recombination studies 200–4
structural parameters *192*
temperature stabilisation 199
SLA analysis, methods comparison *112*
SLA modelling 120–1
  multi-cavity structure by TMM 124
  by one-dimensional waveguide 121–5
SLD, as amplifiers 44–8; *see also* heterostructure; homojunction
span of transmission 3
spectral energy density 13
split contact laser amplifier *52*, 53
spontaneous emission 10, 55–6
  amplifier characteristics 126
  average 117
  beats 153, 154, 155, 156
  effect on SLA 149–50
  fluctuations, and material gain coefficient 169
  gain saturation 112
  intensity, matrix methods 117
  modelling 160–2
  noise 20–2, 149; *see also* shot noise
  noise photons, FPA 153–4
  noise power 28
  non-radiative recombination 39
  phase fluctuation 25
  photons 115, 203–4
  reflection-enhanced *172*, 173
  stray reflections 170–5
  suppression *175*, 177–8
  and TMM analysis 170
  two-level system 11
spontaneous emission factor 142
spontaneous emission power
  Green function approach 183–6
  laser physics analysis 162
  modelled 172
spontaneous emission probability 109
spontaneous emission rate 13, 54
spontaneous emission recombination rate 55
stimulated emission 53–5
  lasing 13–14
  material gain coefficient 37
  population inversion 36, 110–11
  recombination equilibrium 111
  two-level system 12
stimulated emission coefficient 109
STM (single transverse mode) 76, 78
  BH SLA 87

stripe geometry TWA, gain *91*
sub-networks, equivalent circuit modelling 163–4
switch
  bistable 59–60
  detuning 59–60
  gating 23, 59
switching, SLA 59–61
system building 139–41, 182–3

TE (transverse electric) mode 70
  field expressions *71*
  noise power *217, 218*
  quasi 76
    EIM solution 78–9
    optical confinement 87, *89*
  solutions 73
  see also hybrid modes
TE mode gain experiments 211, *212*
temperature
  heterostructure gain 47
  negative 16
  peak gain clamping 43
  resonance control 107
temperature control, experimental laser 194–7, 199
thermal equilibrium, and amplification 13
thermal equilibrium population density 13
TLM (transmission line matrix) *119*, 119–20
TM (transverse magnetic) mode 70
  field expressions *71*
  noise power *217, 218*
  quasi 76
    EIM solution 79–80
    optical confinement 87, *89*
  solutions 73
  see also hybrid modes
TM mode gain experiments 211
TMM (Transfer Matrix Method) *119*, 120
  amplifier structure 135–9
  AVPD compared 129–30, 214
  computation 127–8
  gain calculation *212, 213*
  longitudinal travelling field analysis 120–5
  multi-cavity structure 124
  non-uniform gain profile SLA 125–7
  optical gain 124–5
  and spontaneous emissions 170
  stray reflection effect 173–5
  for travelling wave equations in cavities 162–3

transfer matrix 115, 143–4
transient response modelling 120
transition 10–11, 53–6, 152; *see also* spontaneous emission; stimulated e.
transition probability 54, 55
transition frequency spread 15
transmission line, lossless 132, 162
transmission medium, and communication
speed xi
transmission rate, SLA *48*
transmission span 3
transmittance
field amplitude 122
reflection modelling 171
transparency carrier density 35
transverse field: *see* modal field
transverse wave numbers 76
travelling field, longitudinal 120–5
travelling wave equations 114–17
assumptions 116
noise formulation 159
in optical cavities 162–3
TWA (travelling-wave amplifier) 17
bandwidth utilisation 24
diffusion coefficient 184–5
electric field 184
FPA compared 50, 178
Green function spontaneous emission power
modelling 184–5
noise power generation *176*
photon statistics approach 110
physical parameters and noise figure 179–81
polarisation insensitive 87–98
active layer width 96–8
layer thickness 90–4

refractive index distribution 94–6
practical 51
uniform gain spectrum 107
twin-guided laser amplifier *52*, 53
two-level system, radiation interaction 10–16

wave equation, scalar 69
wave equation solution: *see* modal field solution
waveband 1–2, 3; *see also* bandwidth
waveguide
advantages xii
aspect ratio 97
bandwidth restriction 18, 19
boundary conditions 75
defects 36
dielectric 36
directional confinement 97
effective index 79
heterostructures 46
multi-cavity, in TMM analysis 121–5
one-dimensional, BH SLA modelling 121,
121–5
propagation discrimination 74
rectangular
design parameters 81
dispersion curves 80–2
modal field solution by approximation 75–
6, *77*, 78
waveguide aspect ratio, dispersion curves *82*
wavelength, peak gain 44; *see also* frequency
wavelength requirement for AVPD 130
waves, evanescent 71
Wronskian 183, 184, 185